Usability and the Mobile Web

ALA TechSource purchases fund advocacy, awareness, and accreditation programs for library professionals worldwide.

lita

Usability and the Mobile Web

A LITA Guide

Junior Tidal

ALA TechSource
An imprint of the American Library Association

CHICAGO 2015

Junior Tidal is the multimedia and web services librarian and assistant professor at the Ursula C. Schwerin Library at the New York City College of Technology, City University of New York. His research interests include mobile web development, usability, web metrics, and information architecture. He has published in *OCLC Systems & Services, Computers in Libraries,* and *code4Lib Journal* and contributed chapters to the books *Web Analytics Strategies for Libraries: A LITA Guide* and *The New Academic Librarian: Essays on Changing Roles and Responsibilities*. He has served as secretary for the Library Association of the City University of New York and in the past cochaired the group's Emerging Technologies Committee. He also cochaired New York City's code4lib chapter from 2013 to 2014. Tidal earned his master's degree in library science and his master's degree in information science from Indiana University.

ISBN: 978-0-8389-1301-7 (paper)

Library of Congress Cataloging-in-Publication Data

Tidal, Junior.
Usability and the mobile web : a LITA guide / Junior Tidal.
pages cm
Includes bibliographical references and index.
ISBN 978-0-8389-1301-7
1. Mobile communication systems—Library applications. 2. Library Web sites—Design. 3. Mobile computing. 4. Web site development. 5. Application software—Development. 6. User interfaces (Computer systems) I. Title.
Z680.5.T53 2015
004.1675—dc23 2014042647

Book design in Berkeley and Avenir. Cover image by ©Shutterstock, Inc.

♾ This paper meets the requirements of ANSI/NISO Z39.48-1992 (Permanence of Paper).

Printed in the United States of America

19 18 17 16 15 5 4 3 2 1

Contents

Preface

Mobile technology has changed the way we interact with one another and our communities. For libraries, it's important to create a positive experience for patrons so they can make repeated returns to our institutions and make use of our services. This is true for both virtual and physical representations of the library. Now, with the recent proliferation of mobile devices and constant availability of information access, it is more important than ever for libraries to meet users on their phones and tablets. This book explores usability within the context of mobile web and applications for libraries.

I found it necessary to write this book as more libraries adopt and develop mobile technologies. From my personal observations, I have seen students in my academic library write term papers on their smartphones and conduct research on their touch-optimized computer tablets. Havelka and Verbovetskaya have stated the importance of its "beginning to look like the days of the desktop and laptop as the preferred ways to search, access, and read information are numbered."[1] This is especially fascinating, since this generation of students does not know a time when the Internet wasn't easily accessible, especially with a handheld mobile device.

Now, instead of possessing a laptop or desktop machine to access the Web, more and more students are using a device that fits in their pocket that has immediate access to information resources, anywhere and at any time. Today's mobile devices have the appropriate power to accomplish many of the common web tasks accomplished on a desktop machine. Using these devices may be more favorable for college students, since they have a lower economic and technical barrier for access. The compact physical characteristics of the device and its inherent portability may also be appealing.

I believe it's necessary for libraries to be in the forefront of mobile web use, so that reliable and accurate resources can get into the hands of students. Librarians and developers must recognize the importance of displaying these resources in a meaningful way that patrons can utilize. Once a technology has value for a user, it's logical to assume that it will have repeated use. However, not all mobile library websites are created equal. Creating a desktop library website can be a huge undertaking with a number of challenges that librarians specifically face when developing mobile websites and apps for patrons.

This book aims to help librarians and library developers create mobile websites and applications from a user-centered perspective. That is, to include the user's outlook and mindset in every aspect of the design process. This ensures that the library's mobile project has value for patrons, based on the data and input gathered from its websites. This book is also intended for librarians who have not conducted mobile web usability tests. Librarians and readers who work in libraries who are new to the concept of user-centered design, user experience, and even initializing their first mobile website may also find this text useful.

The difficult part about this mobile usability guide is that technology can outpace the currency of this text. It has been quite a challenge to keep track of mobile usability, since the technology changes so much. Even within a span of a few months, a new device or app has been released, completely changing the way we interact with our devices. This can change how users interact with technology, and greatly impact how testing can take place. This is one of the reasons why mobile usability testing differs from desktop testing. It's commonplace that libraries and librarians have conducted web usability testing on desktop machines, but mobile usability differs significantly from these practices. Mobile users have a myriad of context, screen sizes, input devices, and network connectivity, which separates them from desktop users. Recognizing these differences in the early stages of the design process of a mobile site can greatly enhance the usability of a website, and may require a different method to evaluate their input in comparison to desktop sites.

As much as there are differences between testing a desktop and a mobile device, such as interactions and entering user input, there could be even greater differences in future, emerging technologies. We are on the forefront of wearable technology, such as Google Glass, smart watches, and fitness trackers. In the future, these devices may also be used to access library resources. It may be possible to apply the foundations of mobile usability testing to future generations of users.

NOTE

1. Stefanie Havelka and Alevtina Verbovetskaya, "Mobile Information Literacy," *College & Resource Libraries News* 73, no. 1 (2012): 22–23.

Acknowledgments

I cannot express enough thanks to the staff and faculty of the Ursula C. Schwerin Library of the New York City College of Technology, CUNY. Special thanks goes to Chief Librarian Professor Darrow Wood for his continued encouragement and support of this project. I would also like to thank Patrick Hogan, editor of this guide, for his thoughtful comments and commentary. His feedback has been indispensable.

The members of New York's code4lib chapter and the Library Association of CUNY Emerging Technologies Committee and Mobile Technology Roundtable should also be thanked for their continued insight into these devices that we use every day. Special thanks go to Monica Berger, Robin Davis, Stephen Francoeur, Stefanie Havelka, Steven Ovadia, Maura Smale, Alevtina Verbovetskaya, and Stephen Zweibel for our numerous conversations about mobile and open-source technologies.

I would also like to thank my parents Angelina and Abubakar for taking me to the library every day. They showed me the value of knowledge and how it can impact not only a lone individual, but a whole community as well.

Last and certainly not least, a special thank you goes to my caring and loving wife Elyse. She has a depth of understanding and tolerance that placed me in awe of her while I was working on this manuscript. Writing a book and planning a wedding at the same time are a massive undertaking, but with her at my side, anything is truly possible. You have my heart-filled thanks.

Chapter One

Mobile Devices

Libraries have embraced mobile devices. Smartphones and tablet computers are frequently used to access library websites and resources. Some of these mobile devices can even be borrowed from libraries, where mobile-friendly electronic readers (e-readers) and electronic books (e-books) are widely available at academic and public libraries. According to the Pew Internet and American Life project, these devices are increasing in popularity.[1] Numerous library databases now have mobile counterparts, such as mobile websites and apps, which provide access to their services. Catalogs and discovery tools are becoming mobile-enabled and optimized. With this popular technology, librarians and library web developers must make their own services accessible to mobile users.

It's apparent that the proliferation of smartphone use and ownership has driven mobile web development. A survey conducted by the Pew Internet Research Center in June 2013 found that 56 percent of 252 sampled Americans identify their cell phone as a smartphone.[2] This is increasing among adolescents, with 1 out of every 5 young Americans accessing library websites, specifically library websites, with mobile devices. Libraries must meet the needs of these mobile users, providing them with a satisfying user experience.

The purpose of this book is to help library administrators, web developers, and other staff involved in the library website integrate a mobile perspective in their thinking about websites' usability. It advocates for and describes the task-testing process. This will help identify users' needs and how to meet them through mobile web and application usability testing through the development process. Mobile web usability can provide insight into numerous questions, such as: How can librarians target users in the context of the mobile Web? What differences are there between a traditional, desktop web experience and that of a mobile one? More importantly, how can we better understand how and why mobile users interact with a library website?

MOBILE CONTEXT

An obvious commonality of mobile devices, including smartphones and tablet computers, is that users are "on the go." The small size of mobile devices allows for easy access and use. Information access not only fits in your pocket, but quite literally, in the palm of your hand. They can be used anywhere and at any time.

The "mobile context" refers to the circumstances and environment in which the mobile device is used. The portability of these devices also changes the context, since the circumstances and environments are always in flux. The mobile context can change rapidly. For example, imagine the mobile context change seamlessly from waiting in line for a movie ticket, with mobile device in hand using a movie app to gather film critic reviews, to checking e-mail in your seat before the film starts. The user is placed in two different contexts, interacting with different sets of people and environmental elements. This can distract the user when using the device. This variable context is a fundamental difference between a mobile user and a desktop user.

CONNECTIVITY

Connectivity and broadband speed can play a significant role in the usability of mobile devices. The lack of a network connection at times dictates the usefulness of a phone. How can a phone access your library website if there is a poor-quality connection? A slow connection creates a poor user experience on desktop machines, but it can completely undermine a mobile device. This can limit mobile applications and make mobile websites inaccessible.

HARDWARE

Zhang and Adipat have listed different aspects of mobile usability that can be summarized as hardware issues.[3] The hardware of handheld devices varies widely. Smartphones, tablet computers, e-readers, MP3 players, and even portable gaming consoles can be used to access the mobile Web. Hardware is not limited to just Apple products, Android/Google products, or Windows devices. Ideally, a mobile website should meet the needs of different hardware configurations and be capable of *cross-platform compatibility.* There are a variety of small screen sizes, screen resolutions, and processing speed that all contribute to the user experience.

Small Screen Size

Probably one of the most apparent hardware differences between a mobile device and desktop machines is the screen size. Adipat and Zhang note how small screen sizes can create usability problems. Websites may not only become unpleasing to the eye but cannot be navigated or legible. Screen size and resolution have made more progress since Adipat and Zhang's 2005 article. Raptis et al. have found that larger mobile screen sizes support certain information-seeking behaviors.[4]

Low Screen Resolution

Screen resolution refers to the number of pixels that are on a screen. A pixel is a measurement of light on a screen. The lower number of pixels there are, the lower the screen resolution of the screen. The screen resolution of a mobile device is much lower than that of a desktop machine. This can affect not only how graphics are displayed, but can make text hyperlinks difficult to tap. Graphic images that use text may also lose fidelity when displayed at a lower resolution.

Processing Power

The processing power of mobile and tablet devices is vastly different from a desktop workstation. Web programming such as JavaScript or Flash may not be necessarily compatible with all mobile devices' web browsers and platforms. However, this may have changed because mobile devices have much more processing power today than they did in 2005. One approach is to design and develop mobile applications that do not utilize too much processing power, since this can crash mobile devices.

Data Entry

Data entry on mobile devices is more commonly used with on-screen touch-interfaces, and this can be a challenge for users' finger dexterity. Links can be tapped and images can be swiped. Gestures such as pinching may be used to zoom in and zoom out of images. These interactions should also be taken into account when developing mobile websites.

A recent mobile framework called PACMAD, or People At the Center of Mobile Application Development, combines the International Organization for Standardization, Nielsen's framework model, and the limitations that Zhang and Adipat have outlined.[5] This model is geared specifically for mobile applications. PACMAD consists of three factors related to the usability of mobile applications: user, task, and context of use.

The user is a self-explanatory factor of PACMAD. It takes into account the end-users, and their previous experience in using a mobile system. During the design of a mobile application, different types of users should be expected. One comparison that can be made is that of an experienced power user, who prefers shortcuts and advanced features, and that of a first-time user who would rather interact with an intuitive and easy-to-utilize interface. The design of the application should support these two extremes.

Task is a factor that refers to what the user is attempting to accomplish. There may be extra features that can be incorporated into the task, at the cost of usability. The more complex the mobile system is, the less usable it becomes. This may be problematic for the common third-party systems that libraries use, such as integrated library systems, electronic databases, and catalogs. Many of these systems utilize numerous processes and steps that burden the end-user.

Context of use refers to the circumstances in which the user will use the application. This is closely related to Adipat and Zhang's mobile context, which defines the user's physical location, other tasks the user may also be performing, and environmental elements that the user is engaged with.

There are several characteristics that PACMAD attributes to usability. These are closely related to Nielsen's own usability framework and include effectiveness, efficiency, satisfaction, learnability, memorability, errors, and cognitive load. One attribute, cognitive load, is directly applicable to mobile application use. It refers to the amount of cognitive processing that the user undertakes as they engage a mobile application. One example of cognitive load is using a mobile application while walking, which can slow the user's pace. This can be simulated in a field-testing environment, which will be further explored in chapter 5.

Zhang and Adipat's mobile context and the PACMAD framework can be useful tools that can differentiate between desktop and mobile environments. When creating questions for data-collecting instruments such as surveys, focus groups, or usability tests, it can be useful to refer to these frameworks. For instance, a survey question may ask if users use mobile devices while watching television. A usability test can take into account the difficulty of typing into a virtual keyboard while testing out a task scenario.

THE MOBILE EXPERIENCE IS NOT THE DESKTOP EXPERIENCE

To better understand mobile web usability, we must examine the differences between viewing and using mobile websites between mobile devices and using desktop workstations. As Zhang and Adipat have mentioned in their framework, hardware and network speed can be a dividing line between using mobile devices and computers. Screen size, resolution, orientation, processing power, connectivity, and touch interfaces can factor into how well a website can be used.

Screen Size

There is an obvious difference between the screen size of mobile devices and desktop monitors. The screen size of current mobile devices range anywhere between 4.5" to 7", while the dimensions of the average desktop monitor are between 20" and 24". Laptops range between 11" to as large as 19". Web page real estate, which refers to the placement of elements on a web page, is greatly constrained on a mobile device. This changes how content is viewed on a site. A large desktop screen allows for a wide swath of real estate. There is ample room for elements such as multi-column layouts, ads, banners, pictures, navigation, sub-navigation, and footers and headers. On a smaller device's screen, elements that are not mobile-optimized may be obscured, incorrectly rendered, or may not even load. The smaller screen may also restrict other typical web page elements within web pages, such as tables or frames.

Resolution

Today, according to the web information site W3Schools, most computer workstation monitors utilize a resolution of 1024 × 768 pixels or higher to view web pages.[6] This allows website developers to have a point of reference to create the correct canvas size for a web page. Mobile devices have a much greater variety of resolutions since there is a greater variation of device types. Typically, developers design for a resolution of 320 × 480 pixels for mobile devices. Since mobile screens are smaller, fonts and images will not look as sharp as they do on larger monitors.

Device Orientation

Planning the design of mobile applications can be further complicated because some smartphones and tablet computers rotate web pages depending on the viewing orientation of the device. This reorientation can swap the width and height of the screen's resolution, which may be adverse to the developer's original design. When viewed on a mobile device in landscape (see figure 1.1) or horizontal orientation, the page may load into the whole width of the screen. Even further, when a website is viewed in portrait or vertical orientation, the device may zoom out of the visible area of the page (see figure 1.2), which can leave text unreadable or change the layout of the site. These are important aspects to be considered since they can radically affect how web page navigation and content are displayed.

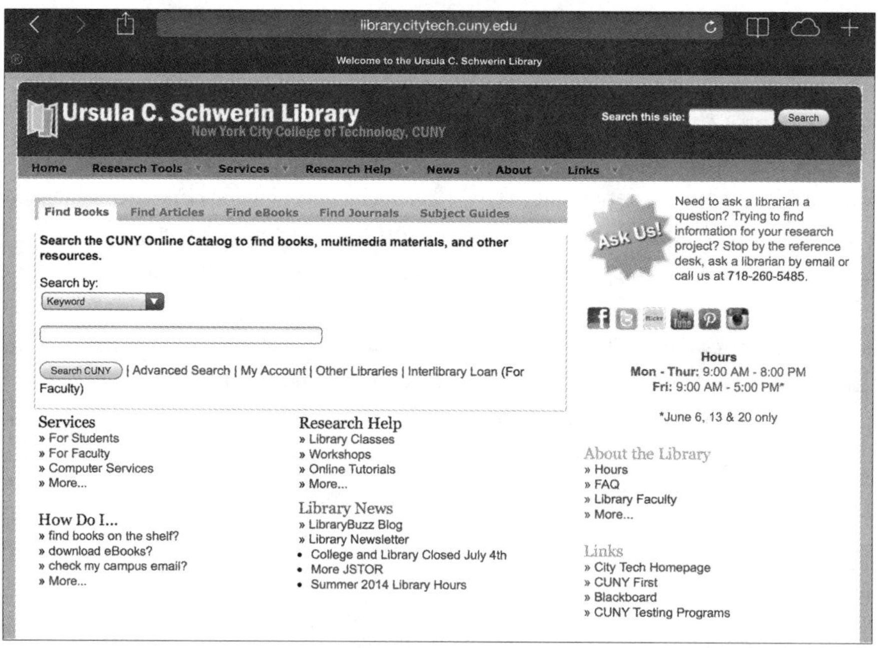

FIGURE 1.1

Website rendered in a tablet's landscape view

Processing Power

Some mobile devices may run on slower processor power than desktop machines. For instance, mobile devices do not have the same multitasking capabilities as does a desktop machine. Multiple windows and tabs can create significant slowdown on tablets and smartphones. Although these types of processors could be seen as a weakness, mobile devices do have some hardware strengths over those of desktop machines.

Today's mobile devices combine and package different technologies together. Smartphones obviously provide voice and data capabilities, but they are also bundled with geo-location capabilities, digital camera, and audio recording. Libraries can potentially harness these features. For instance, the geo-location technology within a smartphone or tablet can potentially be used to display nearby library

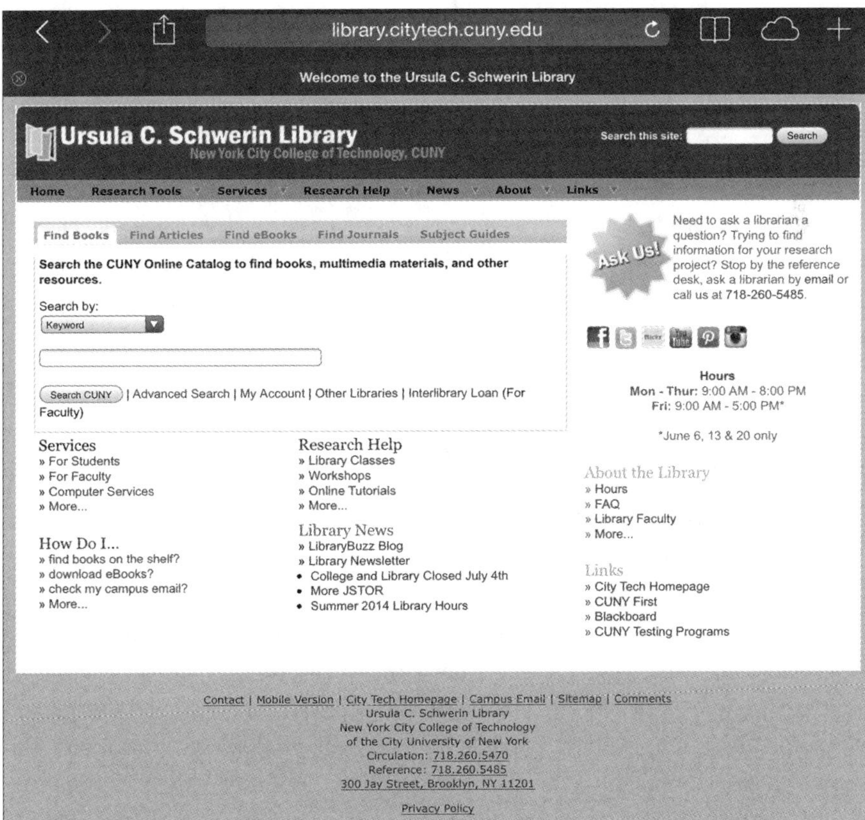

FIGURE 1.2
Website rendered in a tablet's portrait view

branches, based on a mobile user's location. Many libraries have also utilized the social media networks that take advantage of these mobile capabilities, such as the popular Instagram app.

Connectivity

Mobile connectivity can be unreliable in comparison to desktop machines. Cellular data networks are at the mercy of environmental factors, and Wi-Fi can be burdened with too many users, resulting in slower network speeds. This is a specific characteristic that separates desktop and mobile devices and which developers should be aware of.

Touch Interface

Another hardware difference includes the touch interface. This is much different than the keyboard and mouse interface of desktop computers, and tactile-based device interactions have moved from "point and click" to "touch and tap." User input can be limited, especially for office or creative applications. Many mobile devices do have voice recognition as another source of input, but it may not always be available or usable, depending on the situation or context. As such, developers must keep in mind these limitations when creating mobile websites. Navigation can be problematic, since fingers are less accurate than using a mouse, and tapping may not always register.

DEBATE OVER THE MOBILE CONTEXT

Despite the differences between mobile and desktop connectivity, there are also those who believe that the idea of the mobile context may not necessarily apply anymore. Hall argues that the "vague" concept of the mobile context is no longer valid due to the assumption that distractions differ between mobile and desktop experiences, the blurring of handheld devices and desktop technology, and that in some parts of the world, the only network connection is a mobile one.[7] Design and technology have also evolved since Zhang and Adipat's 2005 study. Wroblewski, known for his approach in "mobile first" design philosophy, has noted that the mobile context should be determined by the set of services and content that are provided, not by whether a particular device (mobile or desktop) can meet a user's end goals.[8] Regardless of the debate over the existence of the mobile context, usability testing can help identify and improve what mobile web services libraries have to offer. Perhaps one approach is to use the different attributes of the mobile context and apply modern mobile design and development to augment a library website.

WRAPPING UP

Mobile web usability is no easy task. It requires libraries to develop gradual and incremental improvements to cater to the needs of visitors, based on user research. One piece of the puzzle is examining the hardware, the context of its use, and network connectivity. Librarians can make some changes that will make their mobile websites more accessible and easier to use. The other piece that librarians must solve is how users are going to use the site. The following chapters of this guide provide insight on how this can be done.

NOTES

1. Stefanie Havelka and Alevtina Verbovetskaya, "Mobile Information Literacy," *College & Resource Libraries News* 73, no. 1 (2012): 22–23.
2. Joanna Brenner, "Pew Internet: Mobile,"_http://pewinternet.org/Commentary/2012/February/Pew-Internet-Mobile.aspx.
3. Dongsong Zhang and Boonlit Adipat, "Challenges, Methodologies, and Issues in the Usability Testing of Mobile Applications," *International Journal of Human-Computer Interaction* 18, no. 3 (July 2005): 293–308; doi: 10.1207/s15327590ijhc1803_3.
4. Dimitrios Raptis, Nikolaos Tselios, Jesper Kjeldskov, and Mikael B. Skov, "Does Size Matter?: Investigating the Impact of Mobile Phone Screen Size on Users' Perceived Usability, Effectiveness and Efficiency," in *Proceedings of the 15th International Conference on Human-Computer Interaction with Mobile Devices and Services (MobileHCI '13),* Munich, Germany, August 27–30, 2013 (New York: ACM), doi: 10.1145/2493190.2493204, 127–36; http://doi.acm.org/10.1145/2493190.2493204.
5. Rachel Harrison, Derek Flood, and David Duce, "Usability of Mobile Applications: Literature Review and Rationale for a New Usability Model," *Journal of Interaction Science* 1, no. 1 (2013): 1; doi: 10.1186/2194 0827 1-1.
6. "Browser Display Statistics," W3Schools, www.w3schools.com/browsers/browsers_display.asp.
7. SuAnne Hall, "Three Reasons We've Outgrown Mobile Context," June 3, 2013, www.uxmatters.com/mt/archives/2013/06/three-reasons-weve-outgrown-mobile-context.php.
8. Luke Wroblewski, "Mobile Context Revisited," www.lukew.com/ff/entry.asp?1333.

Chapter Two

Mobile Apps, Websites, and Hybrids

Mobile applications and mobile websites share many characteristics and key differences. It is important for librarians to know these different types so they can make an informed decision on what would be best for their users and institutions. Knowing who your users are factors into the initial and continual success of deploying your mobile project.

WHAT'S THE DIFFERENCE?

Mobile Applications

Mobile apps are software that has been coded for a specific device, which can then be downloaded and installed on that device. The term "app" is short for "application software." These are sometimes labeled as a "native app," because they are programmed and used specifically for that particular device. Throughout this chapter, the terms *mobile app* and *native app* will be used interchangeably. Some of these apps come preinstalled on your device or can be downloaded from the device's specific app store. App data can be stored locally onto the device or stored online for later retrieval. An example of a mobile application may be the camera app built into a smartphone. Both iOS (iPhones, iPads, and iPods) and Android devices are coded specifically for that device.

Apps offer a number of advantages for mobile projects. First, they are optimized specifically for the device they are coded, giving them more processing power. This allows them to utilize the technology inherent within the device. This includes the camera, microphone, accelerometer, and gyroscope. The device's processing power also allows apps to conduct complex functions, such as data manipulation. Native apps are faster than mobile websites since they use onboard processing.

Apps may also allow users to personalize downloaded apps. For example, users can use the onboard device camera to take a picture and use it as a profile picture in an app. This gives the user a sense of ownership. Other personalization and options can also give a more tailored experience for specific audiences.

Networks do not limit apps, since they do not necessarily require a connection. They are not constrained by bandwidth, as images and effects can be packaged with the initial installation of the mobile app. Since they may not require a connection and they have more processing power, interfaces can be more dynamic. Special effects, such as transitions based on user feedback, can also be incorporated into mobile apps.

Mobile Websites

A mobile website is an Internet-accessible website that has been optimized for mobile devices (see figure 2.1). Unlike apps, these are hypertext markup language (HTML) encoded and are accessible through a mobile browser or desktop computer. Just like any other HTML web page, mobile sites can contain content such as text, images, or video.

Figure 2.1 is a screenshot of the Ursula C. Schwerin Library's news blog, *LibraryBuzz*. The site is viewed using the Chrome Browser on an iPhone 4S. The site is powered using the

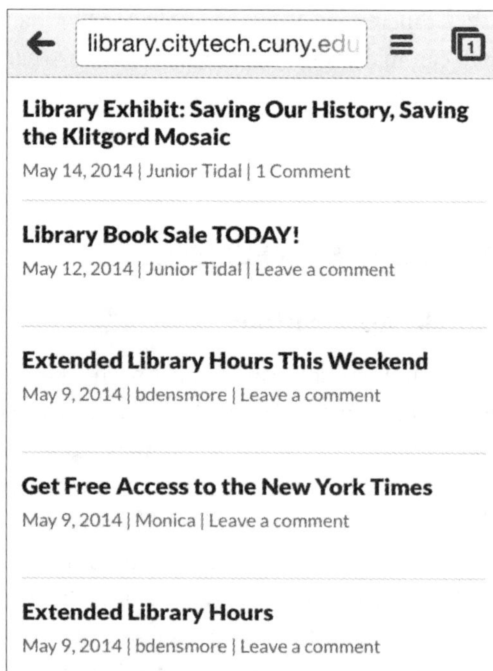

FIGURE 2.1
A screenshot of the City Tech library's blog, *LibraryBuzz*, optimized for mobile devices.

WordPress CMS and the mobile friendly-theme, Twenty Fourteen. The spacing between blog posts is such that they can easily be accessed using touch-based input.

There are a number of strengths that mobile websites possess. Since mobile websites can be coded and deployed easily, they are viewable across a wide range of devices and browsers, which is also known as cross-platform compatibility. Users can visit the website regardless of what device they are viewing it on. Users can access a mobile website linked through search engines since they are indexable. This allows mobile websites to have a broader reach in comparison to mobile apps, which may not necessarily be indexed. Similarly, mobile URLs are more easily shared than mobile apps. A link can be sent via text message, e-mail, social media networks, or in print.

Mobile websites are coded using HTML, which has a low learning curve. This makes them more cost-effective to develop in terms of time and budget. Content is also easy to update. Once a mobile site has been updated, the content is available immediately. HTML also provides multiple design options and frameworks for mobile website creation, which will be discussed later. Unlike mobile applications, websites do not require installation and are ready for launch after coding.

There are a number of drawbacks to mobile sites. Despite cross-platform compatibility, some sites may not render consistently across all devices and browsers. A site viewed in Google Chrome on an Android device may not look exactly the same on an iPhone using the Safari browser. Mobile websites also require a network connection for access. Even if a site is cached or takes advantage of the Application Cache (discussed later), the mobile site needs an initial visit over a network connection.

Content on a mobile site is also constrained by bandwidth. Desktop bandwidth has made site and image downloading an afterthought. Yet, despite their widespread use, mobile devices are at the mercy of network bottlenecks and Wi-Fi signals. This can also impede the static interface of a mobile website, which may have slow responding feedback.

Apps too do not come without disadvantages. Since each device is unique, even among the same operating systems, mobile apps require native development for each device. An app for Android cannot be used on an iPhone, for example, so there may be multiple versions of the same app for multiple platforms. Since these platforms are all different, developing and deploying apps can be expensive and time-consuming. These apps also require distribution approval for their respective marketplaces. This can also delay deployment. This may also apply to subsequent updates to the app. Apps may require even more time to develop and maintain due to operating system updates. For instance, an app may require an update if the device operating system is upgraded. Some apps may not even be fully compatible with upgrades and thus be unusable.

Unlike mobile websites, users must go through a series of steps to use an app. They must search for the app, download it, and install it. This process can make it more difficult to find in comparison to a mobile website. Additionally, URL links to mobile apps can be more difficult to share in comparison to mobile URLs. An iOS app may link to the Apple App Store, but may require users to install the App Store on a nonnative system before they can download it.

Hybrid Apps

To further complicate the mobile landscape, there also exist hybrid web applications. Hybrid apps are a combination of a mobile website and an app. These are built using HTML, JavaScript, and CSS (cascading style sheets). Hybrid apps utilize the onboard web browser engine packaged within the device. It is basically an embedded mobile website inside of an app. Instead of opening a mobile website through the browser, it is done using the browser engine, rendering the web content within the app itself.

RESPONSIVE DESIGN

Others argue that the mobile and desktop experience should be one and the same and that the site should provide a consistent experience no matter what device it's being viewed on. Regardless if the device is viewed on a desktop workstation, a tablet, or a smartphone, the site possesses all of its services' functionality. This is known as *responsive design*. The site automatically adjusts and "responds" based on how it is being viewed by loading the appropriate style sheets depending on the device. The goal of responsive design is to ensure a consistent user experience across many devices. The following images (see figures 2.2–2.5) show the same site through different browser widths. This simulates how responsive design can provide a unifying experience for a variety of screen sizes.

If your library utilizes a content management system (CMS) such as Drupal or Wordpress, there are options to create a responsively designed site. Drupal has plug-ins for various web frameworks such as Bootstrap and HTML5 Boilerplate.

FIGURE 2.2

A screenshot of the City Tech Library Blog viewed on a desktop at maximum screen width, such as a desktop workstation.

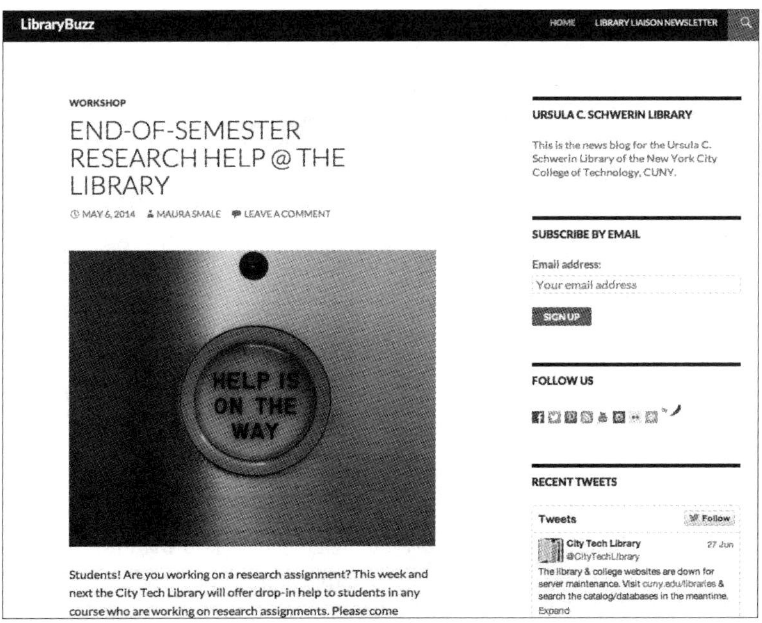

FIGURE 2.3

A screenshot of the City Tech Library Blog when viewed at a smaller screen width. Note the omission of the left sidebar in response to the smaller width.

FIGURE 2.4
A screenshot of the City Tech Library Blog viewed at a minimal screen width, such as a smartphone. A "hamburger"-type menu has replaced the navigation.

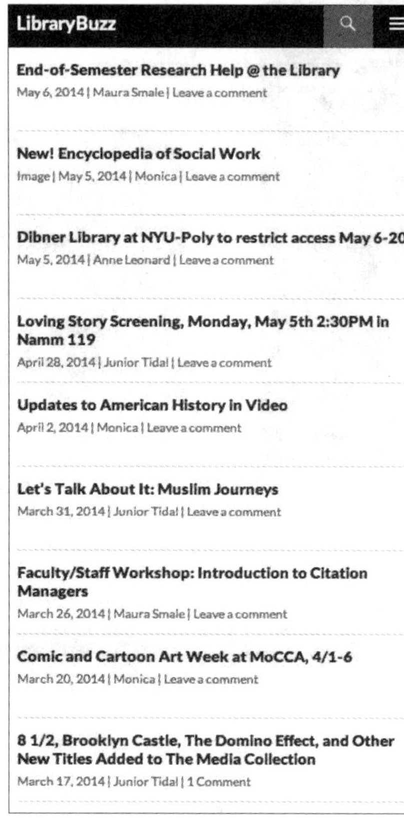

FIGURE 2.5
A screenshot of the City Tech Library Blog viewed at the smallest screen width, displaying only text for faster page loading.

WordPress may also have plug-ins that can make these types of sites mobile-friendly. The packaged "Twenty Fourteen" theme that is built into WordPress is responsive out of the box. There are also many themes for Drupal that are mobile optimized. This is advantageous, since it does not require a whole re-coding of a site, yet may require the appropriate upgrades to the CMS core. Unfortunately, not all plug-ins and themes are compatible with one another, so a responsive design may not be an all-in-one solution for making a website mobile-friendly.

SEPARATE SITES OR RESPONSIVE DESIGN?

In order to balance what can be accomplished through a desktop site and a mobile website, it's important to recognize the tasks and goals that mobile users are seeking to accomplish. Perhaps these goals imply separate mobile and desktop sites. Some designers and developers suggest a strategy where there should be two separate mobile and desktop sites. This means tailoring a specific experience for the appropriate device. A separate mobile website may have some, but not all, of the functionality of a desktop version. For example, a mobile library website may consist of a listing of electronic databases, but may omit those that are not mobile-optimized.

There may exist a better compromise between the debate of a separate site and a responsive one. By determining what resources are available and understanding your users' behavior, you can implement a well-informed mobile strategy. Perhaps it's not possible to code and engineer entirely two separate sites due to budget and time constraints. It may also be possible that a current library website cannot be converted into a responsive website due to the limitations of your current technology or resources. Knowing your users' technological know-how and device capabilities can also factor into this decision.

A good place to start for an initial mobile design is to see what users are doing on your site. Recognizing the basic functions that mobile users wish to accomplish when visiting your library website is one step that can determine an appropriate mobile strategy. Optimizing and simplifying tasks commonly done on the desktop site can help develop a library's mobile site. There are other ways that you can make a mobile website more usable by adjusting for the differences between the mobile and desktop experience.

The smaller screen sizes of phones and tablets can affect the navigation of a web page. One approach is to keep navigation consistent at the top of every page of a site. It is easy to implement and may not require extra JavaScript or CSS. Limiting the number of links is another example of how to make a mobile web page easier to use on a mobile device. This can minimize the number of clicks or taps used on a mobile site's navigation. As mentioned, touch interfaces can be difficult to use and users can accidentally tap a link they didn't mean to use. Placing padding around navigation labels and links may also make it easier for users to tap on what they intended to tap on. Navigation may also be better presented vertically rather than horizontally. The example below is using the jQuery Mobile framework (see figure 2.6). This framework, along with others, will be discussed later. Here, stacking

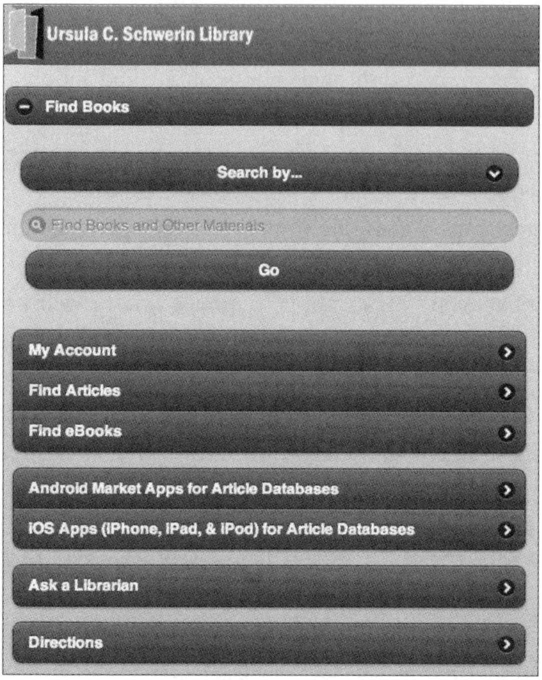

FIGURE 2.6
A screenshot of a mobile website utilizing the jQuery Mobile framework. Note the large buttons and increased font size for touch optimization.

the navigation and reading it from top to bottom is easier than having it run across the top of the page. This can be even more powerful by incorporating nested menus, which collapse or expand into submenus when tapped.

The smaller resolution and screen size of mobile devices require a number of changes toward approaching web usability. One strategy is to reduce content. Since there is less room to display content, text is prioritized or rewritten. With this in mind, text for the mobile Web is characteristically succinct, as users may seek "on the go" information. Users may not necessarily want to, or have the time or context, to scroll through long pages of text in order to meet their information needs. Reducing content on a library website may be a challenge, since these types of sites can provide a wide range of information. Using a focused approach to what information and features are most valued by library patrons is a consideration to make when developing content for smaller mobile screens.

Less content also results in smaller file sizes. This in turn, creates a faster-loading website. This can compensate for slower Wi-Fi signals and cellular data network connections. However, loading large image files on a mobile device can test a user's patience. Reducing unnecessary multimedia, such as removing audio and video, and reducing text and images can greatly support mobile users.

To further optimize mobile websites, developers can "minify" CSS and JavaScript files. This reduces the white space within the file, yet allows it to still be functional when processed by the browser. By removing the white space, the overall file size of CSS and JavaScript files is reduced. This is another method to lighten the overall bandwidth, which can make a mobile site faster to load.

The context in which these devices are used defines and separates mobile and desktop use. Imagine the "desktop context," where a workstation is typically used on a desk with wired connections. The user is restricted to one area to view and utilize the Web. Due to their portability, mobile devices can be always accessible, anywhere, and at any time. Mobile users may also be more likely to have a persistent connection to the Internet through their devices in comparison to desktop users. Despite this, mobile users are at the mercy of a stable Wi-Fi connection and the constraints of their device's battery life.

MOBILE FIRST

The "mobile first" design philosophy takes into account the different characteristics of mobile devices and places them at a primary priority in design. Conceived by Luke Wroblewski, this approach prioritizes the tasks, content, and user experience of mobile users and the development and designing for their devices before creating the desktop version.[1] Smaller screen size, resolution, and hardware are the initial concepts of design. Even though font type and size, screen layout, and interfaces are aesthetic elements that must be altered, the larger issue is content. As a result, a fast-loading mobile site will also load much more quickly on a desktop workstation. The content of a mobile-first design is stripped down to its core essentials. Only the base functionality of a web browser is used and nonmobile features and designs are severely restricted. It should be noted that this does make "mobile first" an unappealing aesthetic.

Responsive Design and Mobile First

The mobile-first philosophy places an emphasis on valued and minimized content. This takes into account the restrictions of a small screen, low resolution, touch-interface, and network dependence, all of which are common among mobile devices. The developer must build from the content up. It also requires the developer to adapt the navigation for mobile devices, possibly restructuring the underlying information architecture of the site.

What Does This Have to Do with Mobile Usability?

The responsive design and mobile-first design are two different philosophies that can be useful for constructing a library's mobile strategy. What makes them useful is that they are inherently mobile-friendly and optimized. These philosophies are also best applied to a single website, rather than creating two separate sites. For example, it wouldn't make much sense to develop a separate mobile website and maintain a responsive desktop site. A single mobile-first or responsive-designed site should be enough to accommodate users of various devices.

Both philosophies, however, have their advantages and disadvantages. A mobile-first experience may be best for a complete redesign of a website, since it places an emphasis on content. It could also be applicable to developing a separate mobile site. As mentioned in the previous chapter, content should be succinct and easy for the user to scan when considering mobile devices. A responsive design might be better to implement for an existing site. Existing CSS files can be edited to include media types and queries to create a responsive site. It may also be easier to maintain and manage a single site rather than two.

A selection of which design type for the user depends on what works best for the user and what resources are available. Developing mobile sites can take up time and resources, but not as much as mobile apps. It's also possible to combine the two philosophies and create a mobile-first responsively designed mobile site. This approach may work best for a complete redesign, since it involves a content inventory and audit. Brad Frost notes that combining the two philosophies supports "designing experiences that work across the entire spectrum of digital devices."[2]

Similarities between Apps and Sites

Mobile apps and mobile websites share a number of similar characteristics. First and foremost, mobile apps and mobile sites are both accessible by a handheld mobile device. Users can download and install a mobile app onto their device or visit a mobile site through the device's network connection. Apps and sites can utilize network access, pulling content from the Internet. Even though a mobile website uses the site's browser, an app can use the device's browser engine to render and display this content.

Mobile sites and mobile apps can also utilize the other technology accessible on a mobile device. One example is accessing a user's geo-location information. Mobile sites can use this to request a user's location and provide data based on this, such as the local weather. One practical application that libraries could implement using this is to provide nearby library branches relevant to the user's location. Even though this is available, access to this technology via a mobile site is greatly limited in comparison to native apps.

Differences among Sites and Apps

Native apps and mobile sites share content, network access, and possess varied levels of access to the features of a device, yet there are a number of differences between them. Network dependence, speed, coding, distribution, hardware, and interface are all qualities that separate the two. Understanding these differences can determine whether an app or a mobile site is best for your particular library institution.

Network Dependence

Mobile websites require a network connection. Mobile apps, on the other hand, do not necessarily need a connection to function. Since a mobile website resides on an external server, an Internet connection over Wi-Fi or cellular data network is required to access that content. If there isn't a network available, the mobile site will not load, or will load using cached data. At times, a network connection may become available but may be oversaturated with other users, resulting in slow, and possibly frustrating, load times. This may impact how website content can be updated on a mobile device.

On a mobile site, content can be updated easily. Developers can simply code or update page content. The user can refresh the page and the content is immediately available. Native apps may have initial content when initially installed, yet require download updates for future updates. Native apps may not necessarily require or use the mobile browser engine to function since data can be stored on the device itself. Once an app is downloaded and installed, it becomes a stand-alone program. Hybrid apps may also have data available for off-line use.

Speed

Native apps typically run faster than mobile websites. Since these apps are coded specifically for their device, there are more resources available to allocate. Mobile websites do not have access to these resources. This allows apps to have animations, video, or audio that loads faster than mobile sites. Like desktop machines, loading time on mobile websites is dependent on the device's network connection speed and the size of the page.

Coding

Mobile websites utilize the same language as desktop web pages for development: HTML. Cascading style sheets are also used. JavaScript is becoming more utilized as more mobile browsers and devices support it. Unlike mobile websites, mobile apps must be coded for their particular operating system (OS). Unlike a mobile site, a desktop workstation cannot natively open a mobile app, unless it's been programmed for that platform. The iOS apps that are used for Apple products such as the iPhone, iPad, and iPod require XCode, Apple's software development kit (SDK), the Objective-C and Cocoa programming languages, and the UIKit framework. Android-based software uses the Android SDK. Blackberry RiM utilizes a variety of languages and programs that include HTML5, the Cascades development framework, Adobe Air, and Java. Microsoft Windows phones utilize Silverlight. Samsung smartphones utilize C++. It is apparent that there is no universal language for mobile app development. Each platform has a specific software language.

Distribution

Since mobile applications are device-specific, they are only available through their respective device or OS vendors. Apple uses their App Store to distribute their applications, Android devices use the Android Market or Google Play, and RiM has the Blackberry App World. In order for users to download apps, they will need to log in to their specific app store, search for the app they are seeking, download the application, and install it on their device. App downloads may require users to purchase the app or make purchases within it. This can be a critical barrier for mobile users.

Mobile websites are not device-specific, and do not need to be purchased, downloaded, or installed. The user can access the site simply by typing in the URL. This makes a mobile website available regardless of the platform or device used. Unlike app purchases, users do not need to enter credentials to access those resources.

The iPhone, iPad, and iPods share the Apple App store and may have separate, specific apps. Android phones and tablets also have separate apps, as do Blackberry RiM-based devices. This means that an app for an iPhone may not work on a Blackberry. If not coded for the platform, the same app for an iPhone may not operate on an iPad. A Blackberry app will not work on an Android device, nor will an Android app work on an iPhone.

Hardware

Since native apps are installed directly on the device, they have the capability to access all the technologies packaged within the device. Mobile apps may use onboard hardware such as its camera, microphone, and speakers. Some apps may also use the device accelerometer, and gyroscope, which monitors if the device is used horizontally or vertically, as well as sense motion. This creates more opportunities for interaction in the app. Apps may also be able to pull data from other software loaded onto the device, such as your address book. A mobile site, in comparison, has limited direct access to a device's hardware or data. Hyperlinks on mobile pages may open up other apps on the device, however, despite the limited interaction between sites and hardware.

Interface

Mobile apps have flexibility with interactive interfaces. They can utilize many of the input gestures mentioned in chapter 1, such as pinching and swipes. These interfaces can also utilize large images or special effects, since they are not constrained by on-demand downloading, like mobile websites. Mobile websites tend to have a static interface. This is slowly changing, however, with the adoption of HTML5 and CSS3, both of which have features specific to mobile web development.

Resources

Budget and time are important factors to consider when developing mobile apps or websites. Mobile sites and application require coding skills and development time. This is not limited to one app either, as a separate app must be created for each individual OS and platform that it is going to be used on. Say, for instance, a library develops an iOS-compatible app. This app would be used only on iOS devices, with the app separately developed for Android devices. Once the app has been coded, it must be submitted for review at each device's corresponding marketplace or distributor. If the project is time-sensitive, this review period can cause an additional delay. Developing and maintaining multiple apps across multiple platforms and distributions can be expensive.

On the other hand, mobile websites can be inexpensively coded. As mentioned before, mobile websites are written using HTML. HTML is a flexible and easy-to-learn language, and requires less time to develop a working product. Source code can easily be extracted from working sites and recycled. Bad code in an app can cause it to crash. HTML is also more forgiving, as the code does not need to be perfectly written for the page to load. Missing tags and empty attributes won't cause a page to crash. Mobile websites also require no approval from vendors for distribution. This makes mobile sites easier to maintain. Once a mobile website has been coded, it can be deployed onto a server and accessed by a wide range of users. However, not every platform and device can render all HTML correctly. This is true for desktop computers as well. There are resources that can be used to determine what HTML and CSS features can be utilized with mobile resources later in this chapter.

Updates

When a mobile app needs to be updated with new content or features, it goes through a process of steps. Apps must be recoded, compiled, uploaded into the app distribution market, and redistributed with the new updates. This requires users to manually download and update their app. Not only are app updates reviewed for marketplace approval, but end-users may choose not to download and update their installed app. This makes app development complicated, since the app, and developers, must service users who may not all have the same versions. This is also true for hybrid mobile applications. Development also does not end after an app has been approved and available for distribution. OS and hardware changes can drive this continuous update of apps.

Developers manage the maintenance and upgrading of apps. OS and device upgrades, or service patches, may require additional testing of the app as time progresses. Major OS updates may also disrupt app development, as features are removed and added for newer platform versions. App testing must be done to ensure that the software is compatible with updated hardware and operating systems. This is very different from mobile websites since they can be updated directly onto the server, with the content readily available.

Which One to Choose?

Choosing between a mobile website, mobile app, or a hybrid can be a challenge. Examining the strengths and weaknesses of each type of mobile application with certain scenarios may assist in choosing what is best to use. Again, compiling this with the mobile strategy of your institution and user research can also help in making a decision. Consider the following scenarios:

A library wants to create a mobile presence. They already have an existing desktop website, but do not have the resources or time to create a whole new site. Perhaps a separate mobile website may be the best way to go. It is inexpensive, easy to code and maintain, and accessible to a wide range of devices. Conducting user research would be helpful to determine how to plan this new mobile, user-centric site. Web analytics can be used to discover what browser, device, or OS patrons are using. Analytics can also be used to determine what content is popular on the site. If patrons are heavily accessing the catalog, finding library hours, and using research databases, then these should definitely be incorporated into the site. Limiting this content can be beneficial, since the small screen size of mobile devices cannot contain everything the desktop site may have and it would make the site easier to update.

Another scenario to consider is implementation of a library's special collection, exhibit, or some other event. A mobile application might fit here best, since there is more flexibility with large image files. Apps are more apt than mobile sites to incorporate media, such as audio and video. The exhibit may have a specified expiration date, which may not necessarily need constant updating. This eliminates the need for releasing revisions or future code and content updates. The app may also take advantage of data from the device's address book contacts to share information. Other features such as geo-location, cameras, or other technologies bundled with a phone may also be accessed. If these features are incorporated in the app, users may want to return and use these technologies frequently. This type of use may also allow the app to be used offline or even off-site. Instead of the patron visiting the library for the exhibit or special collection, they can simply use their phone. If the project is lucky enough to have grant-based funding, perhaps there are enough resources to hire multiple app developers. That way, this application can be available to a wide range of devices.

If institutional resources are available, perhaps it is appropriate to develop both a mobile app and a mobile website. Not only will your institution have a searchable and indexable mobile site, but a mobile app can provide offline services to your patrons. An app and a mobile website will also be able to serve a larger audience. The mobile site can also link to the mobile app and vice versa. The drawback is that both the site and app will require separate updates, which can be costly. A hybrid app, on the other hand, can be a viable solution to circumvent constantly updating two different mobile projects.

NOTES

1. Luke Wroblewski, "Mobile First," 2009, www.lukew.com/ff/entry.asp?933.
2. Brad Frost, "Mobile-First Responsive Web Design," http://bradfrostweb.com/blog/web/mobile-first-responsive-web-design.

Chapter Three

Developing Mobile Websites

Mobile website development is similar to developing a typical website. The bare bones tools necessary for mobile web development include a text editor, knowledge of HTML, cascading style sheets (CSS), and a web browser. The web browser can be the one available on the mobile device, yet there are also a number of desktop emulators that may alternatively be used. For testing mobile usability, it's better to test a mobile website using the actual device rather than an emulator on a desktop machine. This way, users can have the same interface and feedback that they would have when using the device in the field.

Later in this chapter, there will be highlights to some of the features and capabilities that HTML5 has to offer specifically for mobile web development. However, this chapter will not teach how to code such mobile websites nor develop mobile apps. Instead, this will provide an overview of the capabilities possible on a mobile device.

CSS FOR MOBILE DEVICES

CSS is a styling language that defines the look and appearance of a web page. Imagine if an HTML web page is a car, then the CSS determines the size, color, and the coat. CSS can be used to handle the formatting of and layout of a web page. The current version of CSS at the time of this writing is CSS3. It is defined between the tags at the beginning of a web page document as a remote file or a `<head>`

embedded set of rules. In-line CSS can be placed within the HTML page's body tags (see code below). Once the CSS is defined, the web browser processes the style sheet and renders the appearance of the web page accordingly.

This is an example of referencing an external cascading style sheet called mobile.css:

```
<head>
<link rel="stylesheet" type="text/css"
   href="mobile.css">
</head>
```

The following example is an embedded or internal cascading style sheet within the head tags of an HTML document.

```
<head>
<style>
body {font-size: 12px;}
p {color: black;}
</style>
</head>
```


The following is an example of CSS embedded inline in the HTML tags.

```
<html>
<body>
<p style="color: red">I am a snippet of red
   text. <p>
</body>
<html>
```

Specific screen resolutions, screen sizes, font sizes, colors, and layouts can be defined by CSS. Particular screen sizes can have a corresponding style sheet. For instance, there could be a specific style sheet tailored for an Android tablet with a particular set of screen dimensions and another for an iPhone screen. This allows flexibility among different devices, so that multiple devices can have a similar look. CSS can be processed differently among browsers, so it may be important to examine how the site loads on different platforms. Some browsers, such as Internet Explorer 8, may require "hacks" to process the CSS correctly.

There are different ways to switch between CSS style sheets. This can be done at the server level where a processor language such as PHP or Perl loads the correct sheet for the corresponding device. If the PHP/Perl script on the server detects a particular OS or browser type (also known as a user-agent), then the appropriate style sheet is loaded. This detection method can also be implemented using JavaScript. Once the CSS is loaded, a media query can be used to target specific device types (see code below).

The following is an example of a mobile cascading style sheet, which includes a media query applied to devices with a width of 380 pixels. The HTML elements within the body tag are set at an inherited width of 400 pixels, such as the elements contained within the <p> tags.

```
@media (max-device-width: 380px){
body{ width: 400px;}
p{ float: none; width: 100%;}
}
```


Introduced in CSS2, media types were used to load CSS style sheets for specific devices. This has long been used to easily create printer-friendly pages. The code below demonstrates a CSS style sheet specific for printing, where HTML elements using the p tag will appear black and white when printed (see code below). This can also be used to change other elements. The font-size of a web page viewed through a screen, for example, can have a different font size when on a printed page.

In this example, the media query CSS attribute is used to define HTML paragraph tags. The rule specifies that elements and content within the p tags should have black text against a white background.

```
@media print{
p {
color: black;
background: white;
}
```

For mobile web development, the *handheld* media type is specific to small screens and mobile devices. This is useful for making a web page that is consistent between a desktop workstation and a tablet computer, which is the foundation of responsive design. For example, there could be a separate style sheet that defines larger fonts specifically for mobile browsers with smaller screens, and a style sheet that defines font size for desktop monitors. The content remains the same, yet the CSS formats it for the appropriate presentation depending on the device (see code below).

The following example references separate style sheets for a printer device and a smartphone, respectively. Here, the style sheet is specified for each particular device in the "media" attribute.

```
<head>
<link rel = "stylesheet" href="printer.css"
   media="print">
<link rel = "stylesheet" href="phone.css"
   media="handheld">
</head>
```

Media types were not widely adopted because many browsers reacted to media types in different ways. Some browsers would ignore the *handheld* media type, or utilize both the *handheld* and standard CSS and CSS2 scripts, which would create unpredictable effects. Commonly, the layout of the page would become jumbled. CSS3 extended the range of media types for mobile browsers and devices by using *media queries*. Media queries are an extension of the media type that allows custom tailoring for specific browser sizes, screen sizes, and resolutions. A mobile device that meets the qualifications of the media query will load the appropriate style sheet. Examine the following HTML declaration found between the web page's `<head>` tags (see code below).

This example references a default CSS file for desktop devices. It also references a separate CSS sheet for tablet computers that possess a width of 480 pixels. The appropriate sheet loads when the site encounters the corresponding device.

```
<head>
<link rel= "stylesheet" href="default.css">
<link rel = "stylesheet" href="tablet.css"
   media="max-device-width: 480px;">
```

Let's break down the above code. In this example, a mobile device that has a maximum screen width of 480 pixels, or the width of an Apple iPhone, will load the tablet.css style sheet. Widths that are greater than 480 pixels, such as a standard monitor, will ignore this style sheet and load the default CSS file.

CSS can be used to format a wide range of attributes of a web page to accommodate mobile optimization. Font sizes can be increased. Images and extraneous widgets, such as sidebars, can be hidden or pushed to the bottom of a web page. *Fluid layouts,* layouts that conform to a screen's resolution and size, can also be defined within a style sheet. This is the basis for responsive designed sites.

Viewport

The viewport metatag refers to the virtual dimensions of a web page, which instructs the browser to zoom in or out of a web page. These dimensions can be coded into the metatag to optimize or limit the default zoom-level magnification of mobile devices (see below). The content attribute of the metatag defines the width and scale of the web page. The example below indicates that the viewport will be constrained to the width of a device of 320 pixels, with user scaling disabled.

Specifying the viewport in HTML:

```
<head>
<title>Viewport Example</title>
<meta name="viewport" content="width=320, initial-
   scale=0.5">
</head>
```

The same effect can also be implemented using CSS and the @viewport declaration. Since a desktop monitor and a small mobile device have different screen dimensions, the viewport can be defined for the appropriate widths of the page. The above example demonstrates that if the screen width is smaller than 320 pixels, the screen will zoom in to that specific dimension. If the screen is more than 250 pixels, the screen will only show the 320-pixel viewport.

If you've ever visited a website using a mobile device and the page appears too small and you need to zoom in, the viewport may not have been optimized (see figures 3.1–3.2). If the website is optimized using the viewport and initial scale attributes, the page will zoom and adjust for the correct width. In this example (see figure 3.2), the metatag from the above code has been used to render the page. Both of these images were taken from a screenshot of the Google Chrome Browser operating on an iPhone 4S, iOS 7. Having the viewport defined can impact different aspects of usability such as readability, content size, scale, and navigation.

CSS is useful for initial mobile design, but it is not foolproof. Media queries are best used for browsers used on more recently developed devices. As of this writing, CSS is not compatible with Internet Explorer 8 or versions below that. Another

FIGURE 3.1
A screenshot example of an optimized HTML page rendered on a smartphone. The viewport metatag instructs the mobile browser to display the page at a screen width of 320 pixels.

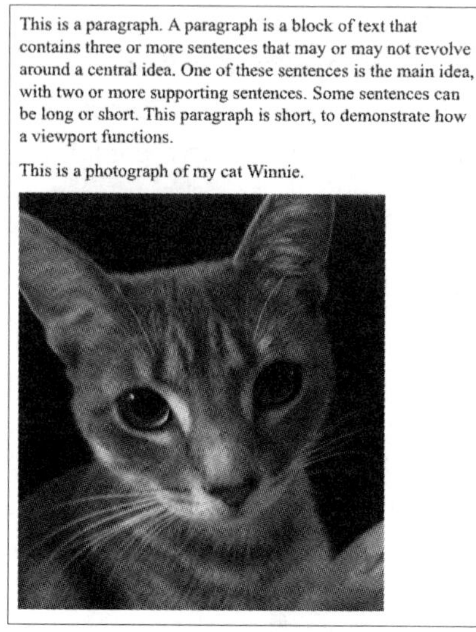

FIGURE 3.2
A screenshot of a page not optimized using the viewport metatag. The text and image are smaller, forcing the user to zoom in.

drawback is the creation of numerous style sheets. Constructing a CSS file for every hardware configuration can be very costly in terms of time and development. One solution is to include one style sheet with rules for separate device widths, yet this also takes time to research and implement. Despite these concerns, the use of CSS media types and queries is fundamental for accommodating websites for mobile devices. It is most notably used in *responsive design*.

HTML5 AND THE MOBILE EXPERIENCE

HTML, the programming language to render web pages, has gone through numerous iterations. HTML5 is specifically optimized for mobile websites and makes up the foundation for mobile web development. This fifth version of HTML was developed to be a seamless experience for most browsers and operating systems,

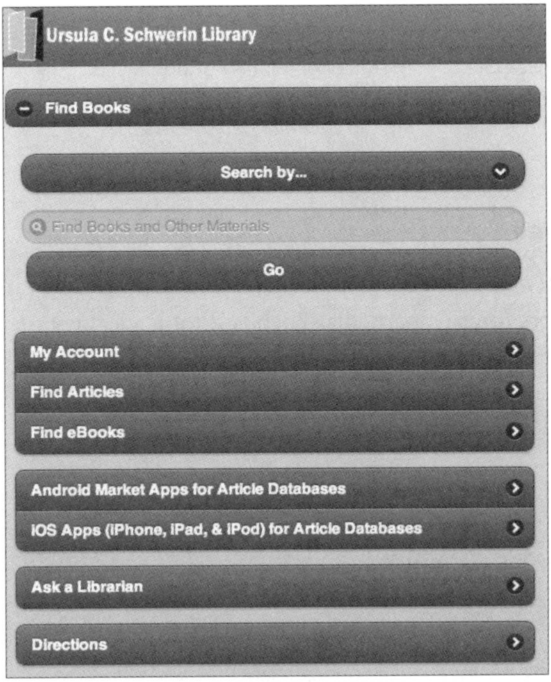

FIGURE 3.3
This is an example of a touch-enabled interface using the jQuery web framework.

specifically with mobile devices in mind. This section reviews some of the features of HTML5 that support a usable mobile experience. There are features that are not fully compatible with all mobile browsers.

HTML5 has a number of features that support mobile websites. This includes touch optimization, usable form elements, off-line storage, and multimedia support. Libraries may not necessarily utilize all of these components, but they do allow for more opportunities to support end-user usability.

Touch Optimization

HTML5 in itself is not touch-optimized, but several web frameworks based on HTML5 are. Sencha Touch, jQuery Mobile, HTML5 Boilerplate, and Bootstrap are some examples of touch-optimized frameworks. This gives mobile web developers many options for interactive, multi-touch interfaces. Some browsers and devices support multiple touch points, also known as multi-touch, on a screen. HTML5 and these frameworks can be used by mobile developers to create large buttons, navigation elements, and other touch areas to prevent imprecise finger taps. The first example uses the jQuery Mobile framework and has large buttons for touch navigation (see figure 3.3). The second example is a non-optimized page viewed on a smartphone that has smaller hyperlinks. Although the second example is coded in HTML5, it doesn't employ a touch-optimized web framework. This can make interacting with the page difficult for users using touch screens (see figure 3.4). Some of these frameworks will be explored later in the chapter.

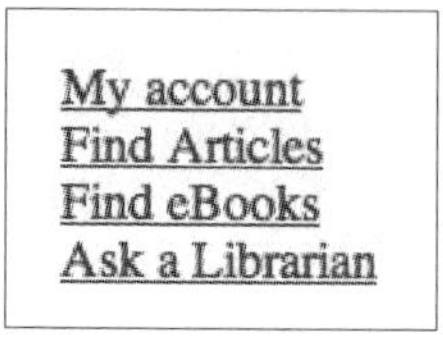

FIGURE 3.4
This is an example of an HTML page that is not optimized for touch interfaces. The links are placed closely together and are a smaller font size. This can potentially cause the user to mis-tap a hyperlink.

Keyboards for Input Types

HTML5 introduces a number of new input types for form elements specifically for mobile devices. This allows users to use a particular keyboard for a specific task. By declaring an input for a particular type, the web page calls for a specialized keyboard to be used. These input types include:

- date
- datetime
- datetime-local
- e-mail
- month
- number
- range
- tel
- time
- url
- week

Different form elements have specific keyboards for mobile input for Android and iOS devices. Using these keyboard types makes it easier for users to input information into a web form. The screenshots below are taken from an iPhone 4S using the Google Chrome browser.

If a form element is tagged with a type of *tel,* which is typically used for entering a phone number, then a 10-digit (0–9) number pad is displayed (see figure 3.5) rather than a full virtual QWERTY keyboard.

The *number* input-type keyboard is similar to the tel-type keyboard, but does not possess the alphanumeric keys commonly seen on a telephone. It also contains the special characters found on the keys above the numbers on a QWERTY keyboard (see figure 3.6).

A keyboard for input type *e-mail* includes the required "@" symbol (see figure 3.7). Keyboards for types of *url* include a forward slash (/), and input requires a valid URI protocol, such as http:// (see figure 3.8).

Date, time, month, and date and time input types use a "roller" for iOS devices, but are not used for Android devices (see figures 3.9–3.11).

Validation types of form elements are used to ensure that the user input corresponds with the correct data. Say, for instance, that you have a Web OPAC that utilizes a 13-digit ISBN to search for books (see code below). The sample code

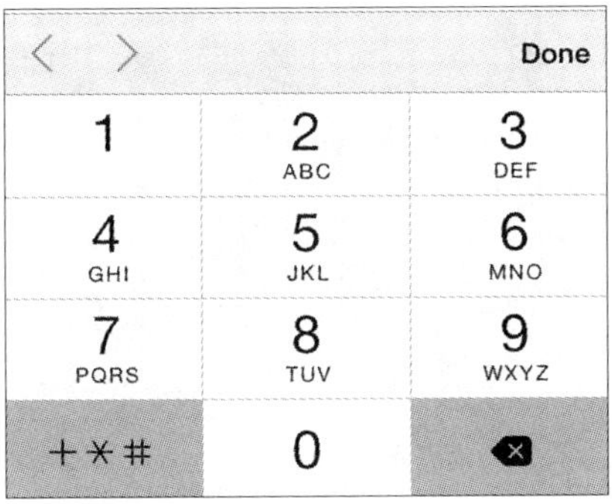

FIGURE 3.5
The tel keyboard displays a telephone keypad. Note the letters mapped to numbers and the addition of the +, *, and # keys.

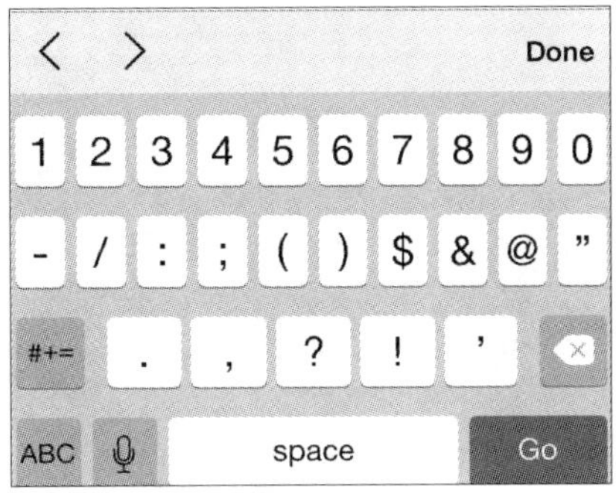

FIGURE 3.6
A screenshot of the numerical keyboard, providing access to number keys and special characters.

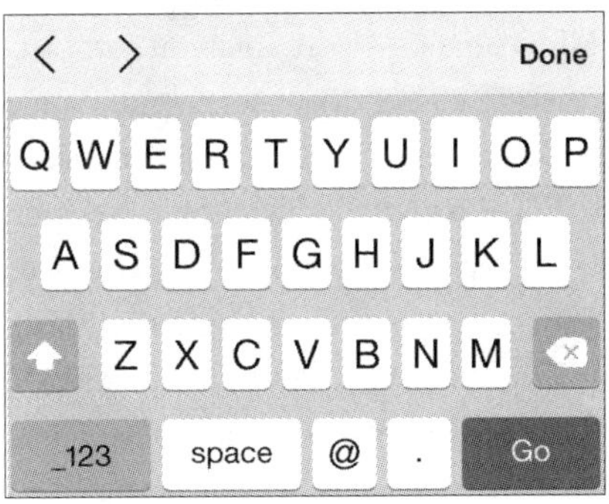

FIGURE 3.7
A screenshot of the QWERTY Keyboard optimized for e-mail input with dedicated @ key.

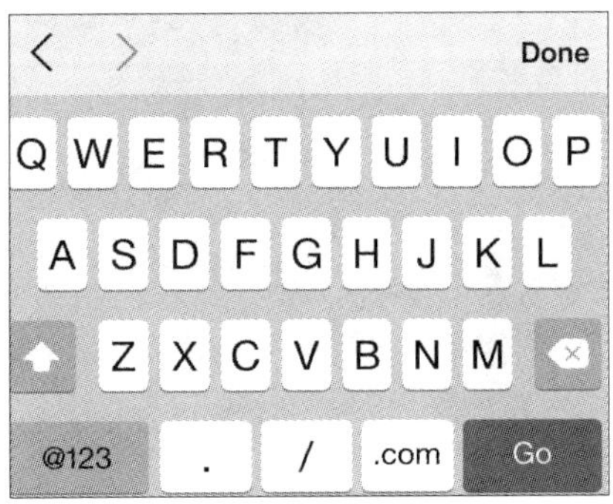

FIGURE 3.8
A screenshot of the QWERTY Keyboard optimized for URLs. Note the / and .com keys.

below can be used to define an HTML input type for specific data. Here, an input type of *number* is declared, where the field only uses numeric values. The *maxlength* attribute indicates that only 13 numbers at most can be entered. The *pattern* attribute is a validation method to ensure that users only type in integers.

The following is an example of an HTML input tag to create a form element for a 10–13-digit ISBN number.

```
<input type="number" maxlength="13" id="ISBN"
   pattern="[0-9]">
```

FIGURE 3.9
A screenshot of a roller input, used for selecting the month and year.

FIGURE 3.10
This roller input is used for a date selection and includes the month, day, and year.

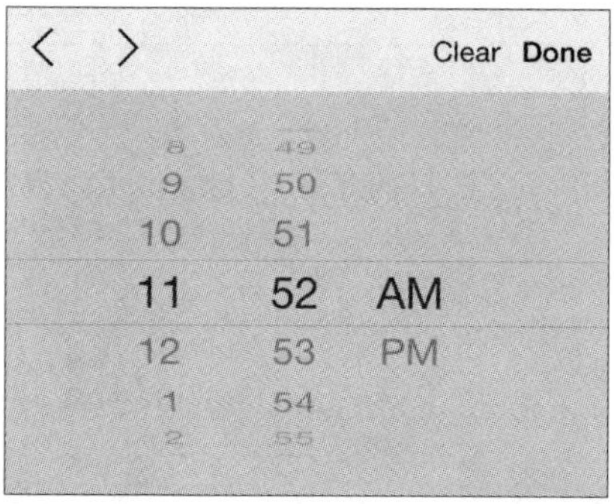

FIGURE 3.11
A screenshot of a time input, which includes the hour, minute, and AM and PM selection, for a 12-hour clock.

These customized keyboard layouts of HTML5 enhances the usability of completing tasks that require user input. Since users do not have the traditional keyboard and mouse, a limited keyboard for specific tasks increases user efficiency. The user no longer needs to fumble with typing numbers on a mobile device, since specific keypads are bigger and easier to use.

Mobile sites can render encoded phone numbers in web pages, so users can simply tap a hyperlinked phone number and the device will dial. This is primarily geared towards smartphones. This can be done using the tel or short messaging system (SMS) protocol for the anchor tag and an "href" attribute in the HTML code (see code below). A device with phone or text-messaging capabilities can access these links and does not require a download or app installation.

```
<a href="tel:555-555-5555">555-555-5555</a>
This HTML tag allows a number to be called when
   the hyperlinked number is activated.
<a href="sms:555-555-5555">555-555-5555</a>
```

This tag allows a number to be sent a text message when the hyperlinked number is activated.

Off-line Storage—The Application Cache

A feature of HTML5 is the Application Cache (or AppCache) manifest. This allows information, such as specific files, to be stored locally on a device without a network connection. This is not necessary when creating mobile sites, but with this enabled, users will be able to view your site offline after an initial visit. Pages will have a faster loading time since local copies are stored on the device for speedier access. Servers will also have a smaller load, since only recently changed or updated data will be downloaded.

The AppCache is declared within the html tags at the top of your web page. Every page that contains this tag will be cached when it is visited. In this example, the myManifestFile.appcache is the manifest file (see code below).

The following is an example of how to reference an application cache manifest file within an HTML document:

```
<html manifest="myManifestFile.appcache">
<head><title>Manifest Cache Example</title></head>
<body>
. . .
</body>
</html>
```

The Application Manifest (myManifestFile.appcache) is a regular text file (see code below). The contents of the file are below and include the *cache manifest, cache, network,* and *fallback* file headers. The cache manifest lists what files should be stored locally. These files are defined under the Cache Manifest section. In this example, the files file.css, file.js, and the image file.jpg will be cached for off-line use. The network declaration is a list of files that require a network connection to be enabled. Here, we have declared two items: the form.html file which will be downloaded every time the site is visited and never cached, and the "*" wildcard. The * wildcard also indicates that no other files, except those under the Cache Manifest header, will be cached. The optional fallback declaration lists files that should be used if a resource is unavailable. In the code above, the trailing slash ("/") matches any URL within the site's domain. If the page cannot be found in the app cache file, the off-line.html file will load in its place. Lines that begin with "#" are comments, which are not executed by the browser. This is useful for developers, since it provides documentation when editing the file and can be used as a way to keep track of the file's time and date. Developers can update accordingly when new content is available.

Below are the contents of a cache manifest. These can be edited using a text editor.

```
CACHE MANIFEST:
```

#Files within the CACHE section will be downloaded to the device and stored locally.

```
/file.css
/file.js
/http://files/images/file.jpg
NETWORK:
```


#Files within this section of the cache manifest require a network availability

```
/form.html
*
FALLBACK:
```

#Files listed in the fallback section are optional URI that are used if the original files are unavailable.

```
//fallback
```

Caching web pages for off-line viewing is not a new concept. It has been used with recent browsers for a number of years on desktop machines. This is exceptional for mobile users, as it can allow access and evergreen information to be present within your mobile site, on or off-line. For instance, library hours or contact information could be referenced within a manifest cache file so that information is accessible to offline users. Again, like many features in HTML5, the app cache may not be necessarily compatible with all platforms and browsers.

Multimedia without the Plug-ins

In addition to the other features of touch optimization and input-specific keyboards, HTML5 code offers limited video and audio support. This allows multimedia to be accessible on the page. This includes the following tags:

canvas—used to draw directly on a web page using user input

audio—used to embed sound or music

embed—container used for plug-ins or other applications

source—indicates the source for <audio> and <video>

track—indicates the track for <audio> and <video>

video—defines video content

Each of these tags has their own specific attributes that are used to control the element on the page. These can also be stylized through CSS and enhanced using JavaScript.

This allows another level of interactivity that was not possible on mobile browsers viewing pages coded using older versions of HTML. HTML5 uses native multimedia, which does not require third-party plug-ins such as Adobe Flash to function. Before HTML5, end-users were required to download plug-ins to view multimedia. Similar to other HTML5 features, multimedia may not be available for every device and browser combination. It should also be noted that this feature should be used sparingly, since multimedia can require a large amount of bandwidth to be delivered.

What Can Be Displayed in HTML5?

Since platforms and browser engines render web content differently, not every HTML5 feature or CSS3 rule can be displayed on a mobile device. Most major browsers packaged with smartphones and a tablet, such as Android Browser and Apple's Safari engine, support most HTML5 and CSS3 features. Two resources that compare these features with various mobile browsers include Maximiliano Firtman's http://mobilehtml5.org/ and the website Can I Use—http://caniuse.com. These sites can help determine what can or cannot be displayed on various mobile browsers, cross-referenced by OS. This may not necessarily be an issue for initial mobile websites, but could prove problematic for more advanced features.

FRAMEWORKS

As mentioned, mobile, touch-optimized web frameworks can be used to develop a site. These frameworks are shortcuts to quickly implement a mobile website. Each of these frameworks has its merits and faults. It is up to the development team's discretion to choose which is appropriate for their mobile project.

jQuery Mobile

There are various web frameworks available that can be used to create mobile websites. One framework is jQuery Mobile. jQuery Mobile is a JavaScript library that facilitates an easier way to develop HTML and JavaScript-based pages for a wide array of mobile devices. This open-source library is free (http://jquerymobile.com/) and is not only cross-platform compatible, but has been optimized for touch interfaces. There are a number of templates that can be used to deliver mobile-optimized web pages. This JavaScript library can be easily deployed locally on a server, or incorporated using a content delivery network (CDN).

A CDN is a system of servers that delivers static content based on the location of the user. It provides on-demand downloads of the latest frameworks, including updates. By including the CDN using the code (see below) in the script tag, web pages can automatically call for the framework to be incorporated into the site without a local server install. These frameworks are also available through their respective websites and the open source software repository Github (www.github.com). The following examples allow developers to link the jQuery Mobile CDN from MaxCDN and Google.com.

Below is a script reference jQuery Mobile within an HTML document. This script commonly goes at the top between the <head> tags, but can be placed anywhere within the <html> tags.

```
<script src="//code.jquery.com/jquery-1.11.0.min.
   js"></script>
<script src="//code.jquery.com/jquery-migrate-
   1.2.1.min.js"></script>
```

The code below is an example of embedding jQuery Mobile from Google's CDN. The script below requests the jQuery Mobile libraries from Google.com. The code is placed within the <head> tags of the HTML document.

```
<link rel="stylesheet" href="//ajax.googleapis
   .com/ajax/libs/jqueryui/1.10.4/themes/
   smoothness/jquery-ui.css" />
<script src="//ajax.googleapis.com/ajax/libs/
   jqueryui/1.10.4/jquery-ui.min.js"></script>
```

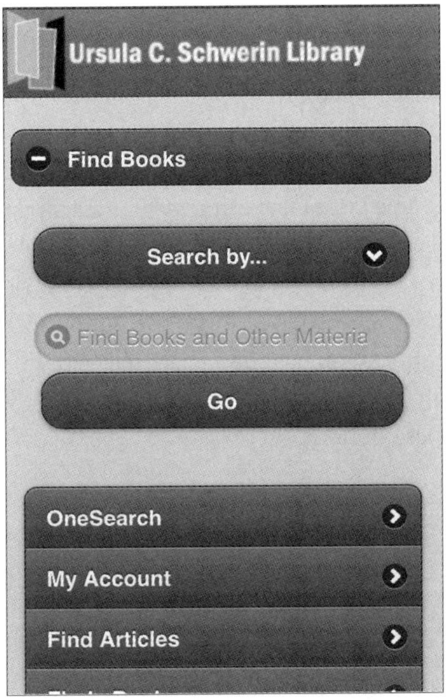

FIGURE 3.12
A screenshot of a jQuery Mobile-enabled library website.

jQuery Mobile has several advantages for initial mobile development. It has a small learning curve, contains a number of free templates, and is lightweight. This particular library is built on the standards of HTML5, so there is less programming involved. This is especially useful, for it allows developers to code a mobile page with minimum programming experience. Additionally, the library has numerous templates that can be built upon or customized. The jQuery Mobile library is also small in file size, which can assist in overcoming the limitations of cellular broadband. The impact that large files can have on the mobile experience was discussed earlier. jQuery Mobile can also be used to incorporate a responsive design. See figure 3.12 for an example of a mobile site powered by jQuery Mobile.

Bootstrap

Formerly known as Twitter Bootstrap, Bootstrap (http://getbootstrap.com) is an open source, fully customizable web framework. Mark Otto and Jacob Thornton, a developer and designer at Twitter, respectively, developed it. Specifically created for front-end development, Bootstrap has a wide range of features including optimized CSS preprocessing using SASS and LESS, a mobile optimized design, and a wide range of documentation and support. Figure 3.13 is an example of what the front-end of Bootstrap would look like in a browser.

What makes Bootstrap a powerful framework is its wide compatibility with a range of devices and browsers. Desktop machines, tablets, and smartphones are able to access websites built using Bootstrap. Its backwards compatibility includes older browsers and current versions of Windows Internet Explorer, in which Bootstrap degrades for maximum compatibility. Bootstrap 2.0 incorporated responsive design practices, yet now uses the mobile first design approach in the latest version 3.0.

Bootstrap also uses a number of components to support mobile website development. These include an image library of "glyphicons" that can be incorporated into Bootstrap's page design. Form controls, labels, and navigation designs are available for use, and are touch- and mobile-friendly. The large form elements, for example, are large to accommodate finger taps and gestures.

Similar to jQuery Mobile, Bootstrap can be installed locally onto a server or referenced using a CDN (see the code below). Bootstrap also has the option of a customized download. CSS files and components can be individually selected. For instance, if a developer doesn't want the glyphicons or navbar components, they can customize a download that removes these options.

Bootstrap CDN

Below is HTML code to reference the Bootstrap CDN. This is linked within the <head> tags of an HTML document.

```
<!—Latest compiled and minified CSS—>
<link rel="stylesheet" href="//netdna.
   bootstrapcdn.com/bootstrap/3.1.1/css/
   bootstrap.min.css">
<!—Optional theme—>
<link rel="stylesheet" href="//netdna.
   bootstrapcdn.com/bootstrap/3.1.1/css/
   bootstrap-theme.min.css">
<!—Latest compiled and minified JavaScript—>
<script src="//netdna.bootstrapcdn.com/
   bootstrap/3.1.1/js/bootstrap.min.js"></script>
```

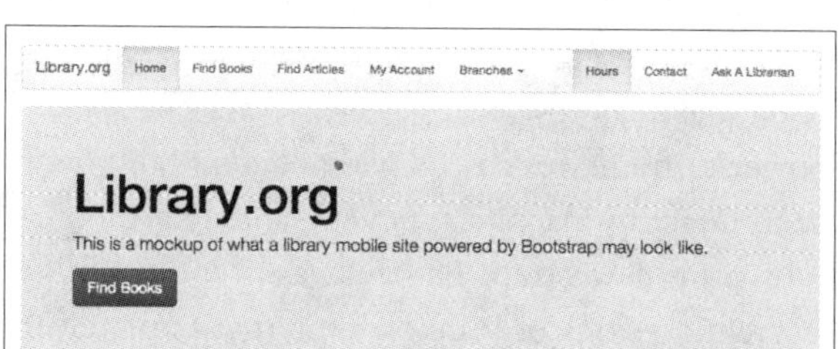

FIGURE 3.13
An example of the Bootstrap front-end framework, which utilizes responsive design.

HTML5 Boilerplate

HTML5 Boilerplate is another mobile-friendly web development kit that can be used for creating websites. Created by Paul Irish and Divya Manian, HTML Boilerplate is an open-source and cross-platform compatible set of templates. Unlike jQuery Mobile and Bootstrap, it is not a framework. It comes packaged with a multitude of elements that can be used "out of the box."

The components in HTML5 Boilerplate include the HTML templates, CSS, JavaScript, an .htaccess file for Apache web server configurations, icons, and other miscellaneous files. The HTML file already incorporates the appropriate doctype in the template. It also includes an optimized print style sheet. The Modernzir JavaScript library is also included in the package. This library detects what browser features are available when the page loads. This can be used to load specific HTML5 resources depending on what browser is used.

HTML5 Boilerplate can also be used in conjunction with Bootstrap. Initialzr (www.initializr.com) is an HTML5 generator, which can be used to create a customized download of an HTML5/Bootstrap template. This template allows the incorporation of Bootstrap components with HTML5 Boilerplate.

DEVELOPING MOBILE APPLICATIONS

As mentioned above, devices using iOS and Android require software development kits (SDKs) to develop native applications. Each SDK has different requirements depending on the systems that they are being developed on. There are other alternatives, such as outsourcing to vendors, to create mobile apps.

Before even beginning to build a mobile app, it may be a good idea to refresh on the user-centered design process discussed in the previous chapter. An app, much like a mobile website, should be thoroughly planned, strategized, designed, and evaluated before launched. As mentioned before, mobile apps take up more resources to execute.

iOS

Native apps for iOS devices (iPhones, iPads, and iPods) require development within a Mac OS X environment. Other requirements include iOS 7 SDK (as of this writing) and XCode 5. Objective-C, a language developed in the early 1980s, is used

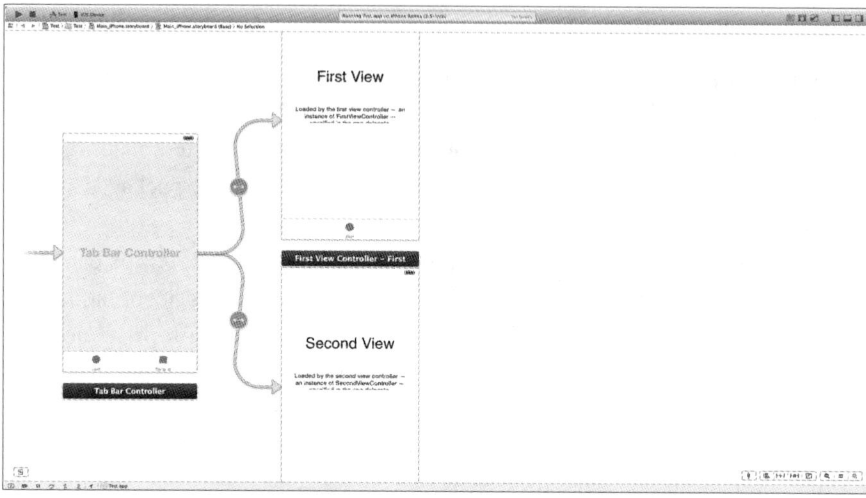

FIGURE 3.14
This is a screenshot of the XCode interface used to build iOS apps.
This screenshot displays the tab bar controller and
the first and second view controllers.

to develop these apps. These tools are freely available through Apple's developer website at https://developer.apple.com. Testing and running the program can be done with the SDK's built-in emulator. Running a native-built iOS app on a device requires a $99 fee to join the iOS developer program and registration with the Apple App Store.

Development within the iOS SDK uses different components to create apps. The integrated development environment (IDE) that the iOS SDK uses is called XCode (see figure 3.14). XCode is available through the Mac App Store for free. XCode uses *storyboards, views, view controllers,* and *navigation controllers.* Storyboards are the containers for the view and navigation controllers. They provide a graphical outline of how the mobile app interface may function. In other words, they provide a layout preview of the app.

A *view* is used to build the user interface of an iOS app. Views have categories that serve different functions. There are several different ones in XCode. The content view provides images or text. The collection displays a group of views. The controls view is comprised of buttons, sliders, and switches. The bars view's function displays toolbars, navigation bars, and the tab bar. The *input view* is used to collect user input. *Containers* are views that enclose other views.

Controllers are the bridge between the user's interactions and the views. The *view controller* is used to interact with the different "scenes" within your app. These are used to render views and pull information onto the screen. If your app utilizes different view schemes, then the navigation controller is used to switch between different types of view controllers.

Just to note, this is just a crash course of how iOS apps are developed. If you are lost in the jargon of storyboards, views, and controllers, don't fret. The Apple Developer Network has a number of tutorials available on their websites (http://developer.apple.com) to help developers get started. These tutorials are also only useful for libraries if the resources are available to create an iOS application.

Android

Native Android apps are built using a variety of tools, which are very different from the iOS SDK. First, Android apps are written using Java. Development is cross-platform compatible and apps can be written under Linux, Mac, or Windows environments. The Android SDK includes the Eclipse Integrated Development Environment (IDE) and the Android Development Tools (ADT) plug-in. Eclipse IDE is the workspace used to develop apps, and the ADT enhances Eclipse for native Android app development. Applications can be freely distributed to devices, yet there is a $24 fee to have them distributed within the Google Play marketplace.

Android apps go through a series of stages in the digital workspace. Similar to the iOS storyboards, the user interface layout is defined. There can be multiple layouts for an app. XML files are used to define the controls and connection between layouts. Image assets and other resources are also used within the development environment. Interactions are defined through Java code. These interactions can be further defined when incorporated into menus and layout transitions.

Java cannot be used on Android devices directly. The Java source code must be exported, then encoded into the Dalvik byte code format (.dex file). Dalvik is the software that runs apps on Android. Once the app has been encoded, it can be freely distributed among devices.

Android app development can be used on these devices or with the Android Virtual Device (AVD) packaged with the Android SDK. Developers can test their apps through different Android versions and configurations using the AVD. It is also possible to define and load different AVD configurations. Developers can then test the app among several different AVD configurations at the same time. AVDs can also incorporate the Google API, so that applications have access to Google Play or the Google maps API.

Outsourcing App Development to Vendors

Besides coding mobile apps in-house, there are also vendors that can develop personalized apps. This includes the company Boopsie, whose apps are targeted specifically for libraries. Boopsie apps implement a number of functions common to mobile library applications. Phonegap is another free app development platform that can use HTML- and CSS-based projects, translating them into apps for a wide variety of devices.

Boopsie

Boopsie (www.boopsie.com) is one example, and provides mobile apps specifically for libraries and their integrated library systems (ILS). Boopsie has many features specific to libraries. This includes a library locator, which uses the GPS of your mobile device. Boopsie also integrates electronic content such as e-books and multimedia, as well as a calendar for events. It also incorporates your existing social media networks with the library app. Boopsie creates apps for various platforms, including Android, iOS, Windows Phone, and Amazon Kindle.

Phonegap

Phonegap (http://phonegap.com) is an Adobe-based development framework that creates mobile hybrid apps that are cross-platform and cross-device compatible. It is free and open source, and uses HTML and JavaScript to build apps. Existing web APIs can also be used to enhance your app. Phonegap also has an online compiling service called build (http://build.phonegap.com), where developers can code the HTML, CSS, and JavaScript files, upload them to the build site, which outputs a compiled version to be reviewed by marketplace distributors.

SUMMARY

Mobile website and app development requires user research and institutional resources to be used properly. Building a mobile web site or app can be challenging to any library, due to the overwhelming number of options that are available.

Knowing your users, the resources available, and the objective of your mobile project can be useful in making evidence-based decisions. There isn't a one-size-fits-all mobile project solution that can solve problems and be useful to your patrons. The next chapter examines how to test your mobile app development through usability testing.

Chapter Four

What Is Usability?

The concept of usability comes in many flavors and definitions. The basic definition of usability is how easy an interface, product, or system is to learn and utilize to meet the goals for end-users. Usability is more common when talking about using websites and interfaces, but it can easily go beyond using a computer. Usability is related to, but is not, user experience (UX). UX is defined as how satisfying and accessible processes can be.

One example of understanding UX is the process of how a patron can go to a circulation desk and check out a book. This is where many steps must be accomplished in order to complete the task. These steps consist of finding the book, taking it to the circulation desk, checking it out under a patron's account, then finally receiving and borrowing the book. This situation may cater to a high level of usability, where a patron can check out a book, but it may not necessarily translate to a satisfying user experience. There could be obstacles that prevent the success of individual steps, which can affect the overall task. Perhaps the user cannot find the book because it was misshelved, or they do not have a library account. These are steps that prevent the user from completing their goal. Identifying the obstacles that prevent task completion is the core of user-centered design. The same idea can be applied to mobile web usability.

To better understand the user experience and usability, researchers have created concept models or *frameworks* that can be applied to better understand user behavior. These models stem from the U.S. Health and Human Services Department, renowned usability expert Jakob Nielsen, and others.

The U.S. Health and Human Services Department defines usability as the quality of a user's experience when interacting with a system.[1] According to usability.gov, this takes into account four characteristics:

- Needs of end-users
- Limitations of end-users
- User preferences
- User's objectives

NEEDS OF END-USERS

Users' previous experiences with a library website, the technology that they are using to view the site, and the tasks they must complete are all tied together. Knowing more about users' backgrounds can be used to serve them better. Recognizing their needs and the reasons why they are on a library's mobile site, what they mean to accomplish, and what they find valuable can be useful to creating a site they will repeatedly utilize. It's not possible to know every aspect of every visitor who comes to a library's mobile site, but there are techniques that can be used to collect information about users. This is known as *user research*. One method to determine more about users is through web analytics.

LIMITATIONS OF END-USERS

User limitations refer to the obstacles and challenges that users may face in using a system. This may include a wide range of challenges. Perhaps a patron may encounter a problem typing in search terms into the online library catalog on a tablet computer. Maybe the home page of your library's website looks strange when using a smartphone. Perhaps users of various levels of technological expertise cannot navigate the library's mobile site.

User limitations can be tied to their previous experiences with a mobile device, visiting the library website, using the mobile web, or a combination thereof. It may not be possible to resolve every usability problem for every individual user visiting your mobile site. Instead, it may be more worthwhile to identify problems that are detrimental to the overall end-user experience. The less there is to think about when using a website, the less frustration for the end-user. The less frustration there is for the user, the more visitors will place value on the library website as an information resource.

USER PREFERENCES

User preferences can help determine what may or may not work in interacting with a website. These characteristics shape the individual experience. One example is how a user opts to browse a library catalog rather than trying to find something specific. This is an important characteristic to take into account, especially in the context of handheld devices. Would the user prefer to scroll endlessly on their smartphone lists of book titles, or would they prefer a grid layout of book covers? Knowing what users prefer can be tricky since everyone uses technology differently. It can also be more difficult to determine what they want if a task can be completed in different ways.

USER'S OBJECTIVES

The user's objective in visiting your site is also a significant factor in user-centered usability. This requires understanding why users are visiting the library's mobile site. User research can help see the purpose or objective a patron is seeking to complete. Most likely, patrons are using the library website to find and gather information, be it about the institution itself or a resource that it offers. By understanding these goals, designers and librarians can make an explicit and clear pathway from the very moment of entering the site to achieving their intended goal. The mobile user, for instance, may have different objectives than a desktop user. Conversely, mobile users may utilize their mobile device much more like a desktop machine, such as conducting deep and time-intensive research.

Another framework of usability is Nielsen's model. Nielsen has stated that usability is a quality attribute relating to how easy something is to use.[2] His model of usability is made up of the following components:

- User Satisfaction
- Efficiency
- Learnability
- Memorability
- Errors

Nielsen's framework breaks down how a website can be understood as usable. User satisfaction can be equated with how positive of an experience the product or system has provided. This is tied into the website's design, but can be arguably related to the user's limitations and preferences.

Efficiency is the measurement of how quickly a website interaction takes for completion. This includes the number of steps it takes to successfully accomplish a process. The fewer steps it takes to perform a task successfully, the higher the site's efficiency. In the mobile context, this can make or break user satisfaction, due to the constraints of the screen size and network speed. Users become frustrated when a site loads slowly. This is especially critical with mobile devices, which may be hindered with unreliable Wi-Fi or strained cellular data networks. One small solution is creating low-overhead and economical mobile web pages that are smaller in file size.

Learnability is how well the user can understand the system when they first use it. The more learnable a system is, the less time it takes to complete tasks. Learnability can also be used to indicate how a user interacts with the system over repeated sessions. A reference librarian using a library website that they interact with every day will probably have a higher level of learnability in comparison with a first-year student who has never conducted research. Simplicity is key to the user's satisfaction with a website. A minimal interface can be more appealing to use than one that has a multitude of features, which gives too many options and can even result in information overload. This is also critical when displayed on mobile devices, with their unique physical constraints.

Memorability rates how a user can return to a previously used system and reuse it. Perhaps a user is trying to find library hours on a website, two weeks after finding it on a previous visit. They remember that the link is somewhere in the upper right-hand corner of not just the home page, but of every subsequent web page on the site. This consistency helps promote memorability, so that users can remember how to interact with the site. An overly complicated website that has numerous links and features may make it difficult for a returning user to remember how to navigate it.

Errors are the mishaps that may occur when interacting with a system, be they user- or system-derived. These errors may include typos, accidentally clicking the wrong link, clicking on something that isn't hyperlinked, or typing in the wrong type of characters into a field, such as numbers into a text box. Some errors may be more serious such as a missing page, lack of system feedback, broken images, or server problems. Web pages will probably have an error one way or another. What is important for the user is the system's ability to prevent errors or to support users to correct them. Users should be able to backtrack from an error.

Nielsen's framework can be summed up simply: maximizing the satisfaction of user through an easily learned, memorable, and efficient system that prevents errors and supports recovery. This results in a highly usable system, be it a mobile website, app, or hybrid. It may require user research and evaluation to understand how and why people use your mobile site. Thinking from the user's point of view is the first step to coming up with a usable design.

These frameworks outline the different characteristics that may be found in utilizing a product or software system. These attributes organize specific concepts to find what does and what doesn't work. It isn't necessary to learn the ins and outs of these different models in creating a usable website, but they are helpful for organizational, development, and testing purposes. The overarching goal of using these models is to make evidence-based decisions from user input so websites can be easier to use. Out of all these, I believe the usability expert and author of *Don't Make Me Think!* Steve Krug, said it best: usability "just means making sure that something works well."[3]

THE IMPORTANCE OF MOBILE USABILITY IN LIBRARIES

Libraries have many stakeholders when it comes to developing a website, mobile or not. Developers, librarians, administrators, and users all have a certain vision and respective goals for the library mobile site, yet they may not be necessarily the same. It's common for these goals to conflict with one another. This is why usability testing is so important. It can help determine what user needs take precedence over those of stakeholders and their institution.

Imagine a scenario where librarians and administrators are tasked with creating a mobile website. The links for library resources and where they should be placed within the website navigation are up for discussion. Should electronic resources be labeled as "databases" or "articles"? Would the casual library patron be able to properly identify electronic resources on the website using one or other of these labels, or would it confuse them? Would they pass up these expensive resources and move on? Usability testing can help answer these questions, which can in turn be used to make informed decisions on users' preferences.

Librarians and administrators certainly have ideas of how the website should function and how it should present information. Yet, they do not have all the answers. If librarians and administrators are not relating to the end-user or lack empathy for the end-user's viewpoint, then the website will create an unsatisfying user experience. For this particular scenario, it is more worthwhile and cost-effective to ask users directly what label makes the most sense to them.

Stakeholders involved in the design of the website may have a different perspective on how things should be. But the people developing the website may not reflect the goals and values of the people actually using the website. Removing obstacles and challenges that block users from reaching their goals and improving their experience are beneficial for everyone involved. Usable design can result in returning patrons, give librarians and developers the freedom to focus on what works, and ensure an effective use of library resources.

A positive mobile user experience can have patrons returning for services. A usable website has consistency and meets user expectations. Providing a straightforward, easy-to-use mobile website allows patrons to find information quickly and efficiently. The easier a website is to learn, the easier it is to use. Users prefer an effortless system, which can result in repeat visits.

For a concrete example of making things easier for the mobile user, consider reducing the length of your mobile URL. Not only does it make it easier to remember, it's easier to enter on small and virtual touch-screen keyboards. For instance, a URL such as http://m.library.org is easier to input than www.library.org/mobile since there are fewer characters to type on a device.

Going a step beyond shortening your mobile URL, another solution is to automatically redirect users. This is known as *browser sniffing,* where a piece of JavaScript or server-side code such as PHP or Perl can be used to detect what type of browser is being used to visit a website. This in turn can be used to serve the mobile version of your library website. If a user is visiting your desktop website on their mobile device it will automatically redirect them to the mobile site. This bypasses a number of steps for the patron to reach your mobile site. This type of development strategy yields a more desirable experience for your end-users. It is always a good idea to include links to a site's full desktop site when on a mobile site and vice versa. There are a number of mobile redirect scripts available on the Internet, as well as plug-ins for different content management systems such as WordPress or Drupal.

Removing usability obstacles not only allows users to easily access the information they seek, but the freedom for librarians to focus on development. Imagine the redesign of a library website specifically to support mobile devices without the input of users who will use the site. Perhaps the library developer who updated the site uses a font that looks great on an Android device, but is too small and unreadable on an iPhone. Hyperlinks on the page may not be activated due to this small size. This shortsightedness can be problematic for non-Android users. Since no usability testing was conducted, there can be no formal feedback on this failure. Not only is the site illegible, but it is also unusable since links cannot be tapped.

This scenario could be avoided if there was some user feedback or information that could be gathered beforehand. Usability can help make informed decisions, and informed decisions can make for better designs. The time and effort to redesign a website can be consuming. Going back and repairing this hypothetical redesign can be wasteful. Having more knowledge about users is more efficient. Starting a website or redesign around a user-centered approach can promote many gains for developers, such as less maintenance and unnecessary coding.

Usability testing is an effective use of resources. One of the benefits of spending time on usability is that it can help make evidence-based, cost-effective decisions. It removes opinion-based decisions and assumptions and opens opportunities to build upon the strengths of a system. These strengths are driven by the expectations of the library's patrons. For example, imagine a library mobile website that goes live without any input from users, based solely on the opinions and whims of librarians. Once it's launched, patrons complain that they cannot find or access the library's catalog from their tablet computer, even though a librarian may find it easy to navigate the website. Labels are presented in complete library jargon and only librarians can use anything on the mobile site. Librarians can find interlibrary loan, the library catalog, serials, and e-journals. Patrons, on the other hand can probably recognize these terms but they would probably not make much sense to them or at worst, hinder their ability to find the information they seek.

The lack of user input not only impedes patrons trying to accomplish their goals, but also carries a high cost. The mobile site is inaccessible with library lingo becoming a barrier, resulting in patrons unable to access the information they need. Placing effort into conducting usability testing can reduce errors and issues. It can certainly save time and money. Even if it's a stand-alone single test for an older site, a usability test can go a long way to save resources. Collecting user input and conducting usability testing can save libraries from the cost of unnecessary repairs or recoding. Understanding the user ahead of time and developing a website are more time- and cost-efficient than releasing a website without user input, only to have ad hoc developers wastefully update errors and repair nonfunctioning aspects.

USER-CENTERED DESIGN AND TASK TESTING

The process of including users' perspectives through each phase of researching, planning, designing, and prototyping refers to *user-centered design*. It is not usability nor is it UX, but it is applying those concepts in a design. The approach is simple: the user comes first. The more a website is intentionally designed for the end-user, the more likely they can accomplish their goals. The easier and less effort a visitor exerts to successfully accomplish their tasks can yield a satisfying experience. Other factors such as institutional goals, features, or technological capability are also taken into account. Yet, every step of the design process places the user as the first priority. Design processes can have various stages, phases, and names. Let's examine a set process that includes planning, analysis, design, and evaluation stages (see figure 4.1).

Evaluation: Formal usability task scenario testing with users can be used to detect issues and errors within the mobile design. Launch and reiterative testing commence.

Planning should consist of user objectives, resources available, scope, and goals for the mobile project.

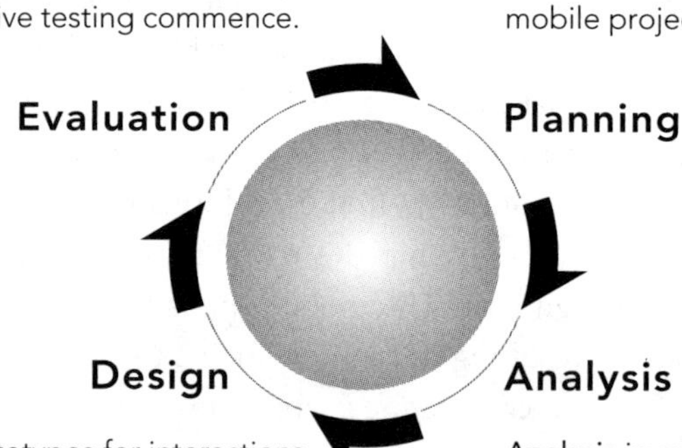

Design: Prototypes for interactions, wire framing for smaller screen sizes, and construction of user-based information architecture from the analysis stage are developed.

Analysis involves user research such as prototyping, or card sorting. Goal-oriented tasks for mobile context can be constructed.

FIGURE 4.1
Sample mobile website design process.

Planning Stage

The planning stage takes into account a strategy for the project, such as determining the objective, resources, and scope of the website. It is the initial stage of the design process, which will set the foundation for the rest of the project. For the purposes of this guide, we are going to imagine that our project is a mobile application for library patrons. The objective determines what the goals of our mobile application are. *Resources* refers to the amount of logical and physical effort that can be placed into the project, such as development and time. The *scope* refers to the library's limitations and requirements. Planning should be focused, where the design is intended for the target audience and is task-oriented. Within the design process, this may mean determining the target audience. For instance, what does a mobile user expect from a mobile library application? Do you need to make every piece of information from your site accessible for a mobile device, or should you focus on the "on the go" type of information and popular content from your site? Another consideration is whether the project is a website and is implementing responsive design practices (more on this later) and how the project accommodates nonmobile users. Again, this should be determined by user expectations, but limited to the obstacles inherent in the mobile web. Surveys and paper prototyping can be used during the planning stages to better understand users. Teams of stakeholders, developers, and usability experts may also be created at this point of the design process.

Analysis Stage

The user's input intersects directly with the analysis stage. At this stage, user research can be conducted to better gather information about and understand the intended audience. This can include the incorporation of analytics information for existing sites. Goals can be identified and established for tasks, which can then be used to dictate the functionality of the mobile site. These goals may be defined by users or developers. It is important to also note that these targets should be universal among all potential users. Once goals are determined for analysis, developers can begin a workflow analysis. This includes the steps and processes that users must undertake to accomplish a task, identified during the planning stages of the design process. These tasks can then be measured and surveyed using usability scenarios and task testing. A content strategy may also come into play, such as making content mobile-optimized, where content should be tailored, and even in some cases constrained, to the expectations of a mobile user. This is also the stage where a requirement analysis, such as obligatory content, of the mobile site will be addressed. Defining the steps can be used to pinpoint potential problems with the design. This stage is completed before the design stage.

Design Stage

The third stage of the design process is called the design stage. The design stage is where a working prototype can be implemented. This involves creating and designing the interactions, also known as wire framing. The information architecture, or a web project's underlying organizational structure, can be constructed at this stage. This can be partially determined from the goals derived from the analysis stage. The site's navigation scheme can then be developed from the information architecture, developing a prototype of the web project. Data collected through user research, such as card sorting, can also be used at this stage to influence the creation of the navigation scheme or to determine a visual hierarchy for the elements within the project. Using this user-generated input, a mock-up of our sample library mobile application can be constructed, either on paper or in working form. The design should conform to the strengths and constraints of a mobile device.

Evaluation Stage

The final stage of the design process is the evaluation stage. Using the goals set in the planning stage, the data and input gathered from the analysis stage, and the development of a prototype, the evaluation stage is used for formal usability testing. Based on the information gathered throughout the design process, modifications to the mobile website are implemented and reiterative testing can be done. The evaluation stage is important because it verifies that the information gathered from users is correctly implemented. Once implemented, the mobile application can be moved from a testing stage to a production environment. Field data can be collected, formal usability tests can be conducted, and any errors can be corrected.

USER RESEARCH

There are a number of different methods to gather qualitative and quantitative data from your users. Usability testing is an observation of how users interact with a system. There are different types of usability testing. Task testing is one method of usability testing in which users' interactions are observed as they are completing a task on the system, website, or mobile application. Analyzing this data can drive subsequent and improved versions of the website. It provides evidence to understand library patrons and not simply assume their needs. Collecting information from task testing also allows librarians and designers to make better evidence-based decisions. The fifth chapter will go into more detail on task testing and how it can be used to collect this data, improving subsequent and iterative designs. Other ways of gaining information include focus groups, surveys, interviews, and a card sort.

Focus groups allow a group of patrons or users to talk about what they like or don't like about a site. A librarian or developer, directly involved with the site development, facilitates these groups. Stakeholders and administrators can optionally observe these focus groups. This type of data gathering is beneficial prior to the design of a mobile website, most notably for initial or redesigned sites, but can be useful for any stage of the design process. It's also important to note that focus groups are no substitute for a usability test. The goal of a focus group is to gather opinions, but it does not necessarily reveal how users will use a website.

There are a number of disadvantages with using a focus group. Focus groups can be expensive in terms of time and costs. Scheduling focus groups and rewarding participants with incentives can be expensive. Facilitators should also be skilled enough to manage a productive session. Even then, the data that is gathered from a focus group may not necessarily be reliable, since group dynamics might encourage the singular opinion of a dominant group member(s). The information gathered might also be inconsistent if followed with a usability test. Focus group participants may explain a process or react to a prototype, yet it may be very different than what they actually do as a participant while completing a task during a usability test.

Interviews are another way you can easily gain information about users. This can be done online or in-person. Interviews can give you information from a personal, one-on-one perspective. They can also be useful for very attentive discussions that may not be possible within groups. Interviews can also be useful for field-testing, which will be discussed later. However, one-on-one interviews can be costly in time and expense. The data collected from interviews may not be usable if the sample size is too small.

Surveys can help find broad similarities and differences among user opinions. They can also be used to find broad generalities among users. Like most research, surveys can be most effective if they have a clear objective or question to answer. Questions should be carefully crafted to get the appropriate response. Open-ended questions can be useful for both interviews and surveys, and can be used as contrast to multiple-choice questions and answers. Multiple-choice questions can be limited in terms of gathering data for user-centered mobile designs. Surveys should also be succinct, since it will take more effort and time to create, release, and analyze them. Chapters 5 and 6 will examine pre- and post-task scenario test surveys in more detail.

A card sort is exactly what it sounds like. A group of users of the site are asked to organize a stack of cards, each labeled with a navigation term used on the site. The user then arranges and sorts the cards to makes the best sense of them. The resulting sorted deck is then used to find commonalities and develop the underlying hierarchy or information architecture of the mobile project. Chapter 5 will revisit card sorting in more detail.

These methods are easy and cheap to implement, with very high gains. Users can be rewarded with small tokens such as candy, library promotional material, or small gift cards for their time. The amount of input amassed from these methods will also be useful throughout the development cycle of the mobile website. Combining user research techniques can be very useful, if they complement one another. For instance, a card sort can be extremely useful to determine what navigation labels work for users. A survey working in conjunction with the card sort can identify what types of users created what kinds of groupings, for instance.

Server logs and analytics can provide some information about users. Server log files are automatically generated reports provided by your web server or hosting provider. These logs collect information such as what type of browser and operating system your visitors are using when accessing your site, the date and time a user has connected to your site, and what page(s) they have visited. This may not provide that much insightful information, but knowing what kind of devices users are connecting to your site with can justify whether a mobile site is necessary or not. Unfortunately, server log files can be problematic to analyze. They are typically created in a raw text format, which need a machine to process and become readable for humans. This is especially true if a site has numerous users visiting the site, as each line of the server log typically represents one visit.

A web log reporting tool can be used to make server logs in a format easier to understand. This may include the popular and free open-source Webalizer (www.webalizer.org) or AWStats analyzer (http://awstats.sourceforge.net). These tools aggregate and summarize log files into human-readable formats, such as tables and graphs. In terms of mobile usability, log-reporting tools are a quick way to see what kind of mobile devices are used to visit your site. This includes what platforms and operating systems are being used. This is useful for librarians who have not conducted user research or implemented online web analytics tools.

The downside of using log files is the large amount of data that can be generated by the server. This is especially true for high-traffic websites. This can also become more complicated if your institution has a number of websites. Each site will typically have its own individual server log. Web reporting tools also take time to analyze and may require periodic interactions to find patterns among users.

Web analytics tools go a step beyond server analysis software. These are usually browser-based web systems, such as the highly popular Google Analytics (www.google.com/analytics) or its open-source alternative Piwik (www.piwik.org), or the many possible built-in reporting tools within your web system, which provide supplemental information. See figures 4.2, 4.3, and 4.4 for screenshots from these tools. The data provided may include your site's most popular pages, referral websites, and what technologies visitors are utilizing to look at your site. All

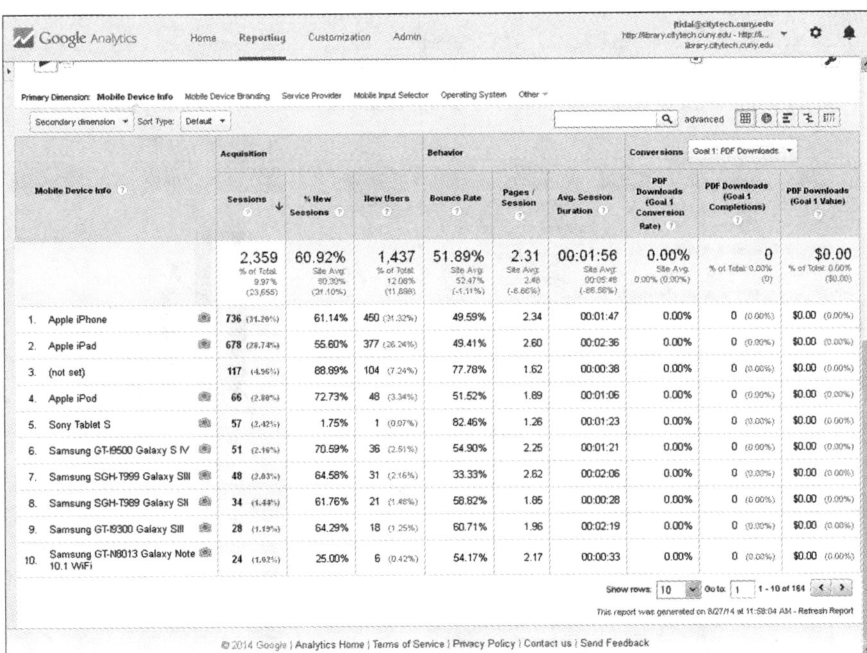

FIGURE 4.2

A screenshot of a Google Analytics page depicting visits from a mobile device.

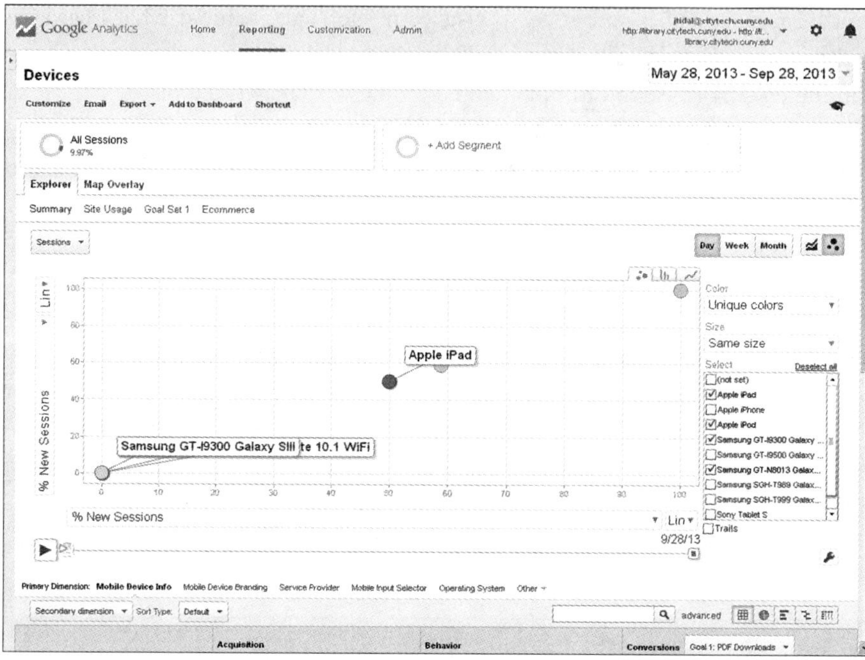

FIGURE 4.3

A screenshot of a Google analytics graph displaying mobile device sessions.

FIGURE 4.4

A screenshot of a Piwik page detailing visitor statistics for both mobile and desktop visits.

of this can be invaluable data, yet it could be difficult to collect otherwise. Much like log files, this only provides a glimpse of what pages your users are visiting in comparison to usability testing.

Analytics have been used at my institution to identify the top interactions of users. For instance, common user behavior includes accessing library hours, using the catalog to search for books, and looking for electronic databases and e-books. Other popular interactions also include asking librarians for reference services or looking for library-specific information, such as phone numbers and e-mail addresses.

Both Piwik and Google Analytics can be used to cross-reference mobile devices with the top pages used on a site. For our initial mobile website, we have used the analytics information of our desktop site to develop the interface. For instance, the hours have been placed at the top of the mobile home page. Links to the catalog, databases, and e-books have been placed in the order of their top usage. This can be useful for initial mobile projects or even redesigns, as they incorporate passive user feedback into the design of the site's interface.

USABILITY ON A MOBILE DEVICE

Web usability testing has traditionally been done on desktop computers. Even though a desktop computer and a mobile device can access a website, there are a number of differences that can change the user's experience between the two. When referring to a mobile device, we are talking about any handheld machine that can access the Web wirelessly, including tablet computers, smartphones, portable gaming consoles, or older personal data assistants (PDAs). Input devices such as a mouse, keyboard, and monitor used with a computer workstation differ greatly from today's modern screen interfaces. Smaller screen sizes, resolutions, network speed, and processor speed can also create numerous differences between workstations and smaller devices. Zhang and Adipat, have created a framework that is unique to mobile application usability:[4]

- Mobile context
- Connectivity
- Small screen size
- Low screen resolution
- Limited processing power
- Data entry

Mobile web usability is no easy task. It requires libraries to develop gradual and incremental improvements to cater to the needs of visitors, based on user research. One piece of the puzzle is examining the hardware, the context of its use, and network connectivity. By examining these, librarians can make some changes that will make their mobile websites more accessible and easier to use. The other piece that librarians must solve is how users are going to use the site.

NOTES

1. "Usability Basics," www.usability.gov/basics/index.html.
2. Jakob Nielsen, *Prioritizing Web Usability* (Berkeley, Calif.: New Riders, 2006).
3. Steve Krug, *Don't Make Me Think!: A Common Sense Approach to Web Usability*, 2nd ed. (Berkeley, CA: New Riders, 2006).
4. Dongsong Zhang, and Boonlit Adipat, "Challenges, Methodologies, and Issues in the Usability Testing of Mobile Applications," *International Journal of Human-Computer Interaction* 18, no. 3 (July 2005): 293–308, doi:10.1207/s15327590ijhc1803_3.

Chapter Five

Mobile Usability Testing

Once you've selected what type of mobile project to use, the next stage is to see how it fares with users through usability testing. Usability testing is an observation of how well a system is utilized by end-users. The main goal is to determine the user's satisfaction with the system and to then improve it through reiterative designing and testing. It does not involve the skill set of the user, but whether or not the system supports users in accomplishing a task. This involves identifying and collecting data on user performance, problems they may encounter internal and external to the system, and retesting them after making modifications to the system. This chapter will look at the different components used to organize, plan, and execute usability tests specifically for mobile devices.

This chapter will also examine how to develop a test plan, prepare for the tasks to test, recruit participants, and conduct the test (see figure 5.1). The analysis of the results from the usability test, reiterative design and development, and retesting will be examined more thoroughly in the next chapter.

DEVELOPING A TEST PLAN

The test plan is the first step in constructing a usability test. At this initial stage, the usability expert or team determines what is being tested, why is it being tested, who is going to test it, what metrics are going to be recorded, where and how the testing will take place, and what tasks are going to be used for the test. The usability

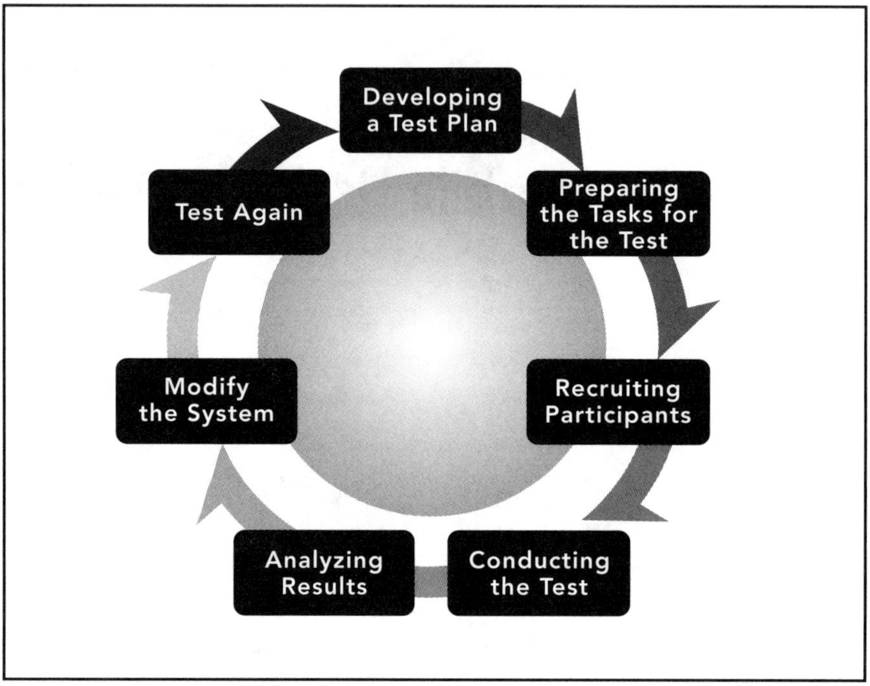

FIGURE 5.1
Stages of a usability plan.

expert or team can be the one in charge of the mobile project, which could be a librarian, a library usability web committee, administrator, or a combination of them. Answering these questions will help concerned parties, such as stakeholders, librarians, developers, testing proctors, and testers understand the goals of the testing. More importantly, it will provide stakeholders a perspective into the user's thinking.

A test plan for a mobile library website can consist of these elements:

> **SCOPE:** The scope refers to what is being tested. This may include specific features or just a certain aspect of the mobile project. The scope also determines what particular stage of development should be tested. This could be the project's initial design, a prototype, or even a working, live project. Developers may also have key questions for participants that can also be included in the scope.
>
> *Example:* The library's first mobile website. It is a working prototype that is not yet live to the public. The main focus of the usability test is finding library materials and the site's navigation.

PURPOSE: The purpose of the usability test is the goals and concerns that stakeholders have. It also determines why testing should take place. These may be general or specific conditions of testing that require feedback.

> *Example:* Are mobile users able to find library materials, such as books, journals, and databases, through the library's mobile website?

PARTICIPANTS: Participants refers to the subject population who will be engaging in the usability test. This also includes how they are recruited and how many to test. Participants should be the potential users of the mobile project. Depending on the scope of the usability test, this could include novice, casual, or power users of mobile devices.

> *Example:* Our participants consist of potential library patrons who use smartphones, tablet computers, and e-readers. Users will be recruited through advertising on the library website, reference desk, social media postings, and in-person recruitment in the library. This usability test will consist of 5 users for the first round of testing and another 5 for the second round.

METRICS: Metrics are the measurements taken during the usability test. Participants' feedback and questions (qualitative) or measurable aspects (quantitative) that are recorded are considered metrics. This will be explored later in the chapter.

> *Example:* The qualitative measurements of this usability test will include responses to how satisfied users are in finding library resources on the mobile site, gathered through open-ended questions and the think-aloud protocol (TAP). Metrics to record include the length of time it takes to complete a task and the number of taps it takes for users to find materials.

TEST LOCATION: The test location is the testing space where usability tests are conducted. This may be a laboratory setting or a test conducted in the field. This will be explored later in this chapter.

> *Example:* Usability testing the library's mobile website will be conducted in the open, public area of the library that has access to a Wi-Fi network.

TEST SESSION AND EQUIPMENT: The test session defines how the usability test is conducted and carried out. Usability proctors should inform participants how long a typical test should last. This should allow time for participants entering and leaving the test session, late participants, and post-test debriefing. Equipment can be provided by either the testing team or provided by the subject. At times, the latter is preferable since not everyone may be familiar with the supplied equipment, which could potentially taint the test. Participants may have a usability problem with using an unfamiliar device, which would be problematic when interacting with the mobile site. There may be other considerations to make with the test session, such as receiving the necessary credentials in using the testing space's Wi-Fi or the participant's cellular data network.

Example: Participants will be informed that testing sessions will last 30–45 minutes each. Participants will bring their own mobile devices to the testing session.

TASKS: These are the tasks and scenarios users will be asked to complete throughout the usability test. These are related to the *purpose* of the usability test. The scenarios will be examined more closely in the next section.

Example: You would like to visit the library, but don't know where it is. Find directions using the library's mobile website through your mobile device.

TYPES OF USABILITY TESTING

There are different types of usability testing that can be used to evaluate a mobile project. Rubin explains that there are four different types of usability testing: exploratory, assessment, validation, and comparison.[1] Each of these types of usability testing has its own merits and methods for developers to use to better understand development throughout the design cycle.

Exploratory Testing

Exploratory testing is used for the early development of an application. This type of testing is useful to see how well intuitive and functional prototype designs are. One example of using an exploratory test is a card sort. This can be used to establish the information architecture of the site.

I have used exploratory card sorting as a method to identify labels for our library's redesign. Library terms and jargon can be confusing to users, so a card sort can be used to identify navigation labels that can effectively communicate to the end-user. For example, in an upcoming site redesign, the label for "catalog" has been changed to the more user-friendly and active phrasing, "Find Books." This can make a significant difference in supporting users' information-seeking behaviors.

Task testing can also be used as a method of exploratory testing if the mobile application has reached a stage of interaction. Paper prototypes would also be another way to conduct exploratory testing, since they are easy and inexpensive to implement. This will be explored later in the chapter.

Assessment Testing

Assessment testing is used for applications, which are in the intermediate stage of development. This goes beyond the initial design and examines how well it has been implemented. Task scenario testing is a method of assessment testing, as it looks at how users interact with the system. This type of testing is largely what this chapter will focus on.

Validation Testing

Another type of usability testing is validation testing. Validation testing measures the benchmark of a system with users. This is intended to be used prior to the launch of an application, later in development. Comparing the timing metrics of a task scenario test with that of a user and an expert is one example of validation testing.

Comparison Testing

Comparison testing is used when there are multiple design versions of an application. This type of testing can be used at any point in the development process, to compare the merits and faults of different implementations. It pinpoints which design version is the most usable. For mobile applications, comparison testing can be used to differentiate the design of a mobile site from that of a desktop one.

USABILITY TESTING METHODS

Card Sorting

Card sorting is an initial usability test that can be used for initial prototypes. In card sorting, a deck of index cards, stick notes, or strips of paper with labels are

FIGURE 5.2
An example card sort using sticky notes instead of index cards.

given for a participant to sort out (see figure 5.2). This usually includes navigation labels used on the website. Participants are expected to group like terms together. A closed card sort has a number of predetermined categories for the user to sort. An open card sort allows the participants to create their own categories or labels using blank cards they can fill in. A developer can analyze the results of the card sort and create generalized ideas about the navigation scheme or information architecture of a mobile website. This usability test is best used among a large sample of participants, representative of the target audience of the mobile application.

Task Testing

Tasks are the *interactions* that are conducted between the mobile project and the user. These can range from the most basic tasks, such as finding library hours, to more complex ones, like finding a particular electronic database. Tasks should be tied to the goals that are defined in the scope and purpose of the usability plan. Task testing is used to find out what problems are inherent in the design of the system. A library mobile website can be well coded and developed, but that does not necessarily imply that it is user-friendly. Even the most simple of websites cannot be free of errors. Task testing helps verify what those problems may be.

Task testing allows user interactions to be observed. Let's look at how a survey question can be compared to a usability task. A question such as "What did you like about the mobile library website?" can be used to prompt the user for an open-ended answer. At times, this kind of feedback can prove to be invaluable, as this information may not be found in visitor numbers or analytics statistics. This is not to imply that a user's opinion is not useful, but there can be more to be learned. By structuring the question as a task, developers can witness how a user carries it out rather than just gather their opinion.

For initial usability testing, compiling the most common tasks performed on the library desktop website is a good place to start. It may be an inefficient use of time attempting to test every task scenario that can be done on a mobile project. Looking at library websites, this may be such interactions as finding a book, searching for the library's hours, or patrons accessing their library account. Identifying five tasks, at the very least, can reveal a great deal of information about how users interact with the site and the problems that they may encounter. Using the same task scenarios across multiple mobile devices may also be used to contrast and compare different platforms and users.

Not only is it important to seek what popular tasks are performed on the library website, but also to examine the essential tasks of the site. Librarians can go use the prototype to predict what aspects of the mobile project are the most cumbersome for the user. Perhaps there is a lengthy process in renewing a book online through the mobile site, and user feedback would be useful to determine the problems inherent in that interaction. It may also be a good idea for the design and development team to go through the mobile project to determine what tasks should be tested. After all, it is the designers and developers who will benefit from user feedback, as they will update the mobile project after the first round of testing.

It is also important to have empathy for end-users of the mobile project. After all, patrons may not have necessarily attended library school, are not information professionals, nor are they web designers or developers. Gathering tasks using this mindset can help usability testers find the appropriate task to test. This may also be necessary to state during the usability testing phase with other librarians, administrators, and developers. For instance, during our last redesign, there was a debate whether to change the label for "databases" to "articles." Although this may make a more accurate label for librarians, end-users may be utterly confused since the term *databases* can have multiple meanings. Usability testing can help identify if this term is acceptable among end-users, which can be used to back arguments and make evidence-based decision making.

A typical usability task test involves a proctor, a participant, the system (mobile site, mobile app, etc.) to be tested, and optional observers watching the usability test. The typical usability process is as follows. A proctor briefs the participant on what the test may entail. This may include an overview of the scope of the test. The proctor reads a script of instructions to the participant. If required, the participant signs a consent form to conduct the test, and to be optionally recorded by audio or video. The test begins when the participant is asked to complete a series of tasks using the system. Participant performance is then recorded while and after they have completed the task. At the conclusion of the test, the proctor has a debriefing session with the participant. Observers may optionally attend the debriefing session, so they can gather more input from the participant.

During task testing, there is a variety of feedback from users that can be recorded. The "think-aloud" protocol (TAP) can be used to gather information as the task progresses. This is where the user is speaking aloud what they are thinking as they are performing each task. This provides perspective from the users' point of view as they attempt tasks on the mobile project. TAP is not foolproof, however. There may be some participants who are reluctant to give their opinion about the system. It is up to the proctor to encourage the participant's TAP reactions so they can be explicit in their responses.

There are other quantitative measurements that can be recorded during a usability test. Probably the most important measurement is whether or not the task attempted failed or succeeded. This is a simple statistic to generate, as it can be a percentage of success among participants. Another metric to include is the number of mouse clicks, or in the case of touch-sensitive devices, "taps" that the user has made during the task. This can be monitored by an observer, second proctor, or through electronic means such as video recording or screen capturing.

Recording the timing of a task can also be useful. As a stand-alone metric, timing may not provide much information. However, the length of time a task takes for a successful or failed completion can be used as a comparison among participants. This may be best measured in minutes and seconds.

Navigation paths may also be another metric to take note of. This a recording of the specific pages a user has "traveled" to complete a task. Perhaps one objective of the usability test is to track the different navigation paths a task may take participants through. This can be used to better understand the efficiency of a task, if there is more than one way to complete it. This may be best observed after the testing session, if it has been digitally recorded via screen cast.

The errors encountered while task testing are another useful metric to record. This doesn't mean counting errors alone, but specifying what errors are encountered, which provides context for developers and observers. User errors such as unintentional taps, straying from the intended navigation path, or selecting the wrong link should be recorded. System errors such as broken links, typos, and missing page or screen errors should also be noted. This will be useful for repairing the mobile project for its next version.

Task metrics can have a baseline metric as a comparison. This could include an expert in using the system, but shouldn't be someone who has direct involvement in the development of the site. A reference librarian could be a potential expert in using the site, for instance. The expert would attempt to complete the task and have their timing and taps recorded. The results can be used as a benchmark for validation or comparison testing.

It is also possible to make a hybrid of quantitative and qualitative measurements. Using a five-point Likert scale, this multi-point scale allows quantifying subjective opinions, attitudes, and preferences either as a positive, negative, or something in between (see below). This can measure how well a mobile website or app is satisfying.

Below is an example usability question using the Likert scale. This scale can be used to measure attitudes, preferences, and opinions regarding a library's mobile site or application.

It was easy to find a book using my smartphone

1 2 3 4 5

Strongly Agree Strongly Disagree

TASK SCENARIOS

Writing appropriate task scenarios is fundamental to task-testing. A task scenario gives usability participants an applicable, real-world scenario and places them in an appropriate frame of mind to attempt the task. It's important that tasks are well worded, so the participant can understand what to do. This goes for the proctor as well, since they may need to be knowledgeable about the tasks in case the participant asks any questions. Here's an example task scenario intended for a desktop website:

"You are a student in an English writing class. You need to find the book *Frankenstein* to complete an assignment. How would you find this book on the library website?"

In this example, the participant is given a hypothetical scenario that places them in a role (a student in an English class) with a specific function to complete (finding a book on the library website). Providing context is more conducive and motivating than commanding users to interact with the site. The task is also succinct, so there is enough information to cue the participant to take an action. Overloading the participant with too many details can also skew the task. The lack of focus of a task can obscure the goal from the participant.

Examine this modified example task scenario:

> You are a student. You need to find a book.

This example gives very little context to the participant. It opens the task for multiple forms of completion and doesn't provide enough information or direction for the participant to use. This is likely to confuse the participant since they could use any book as a viable solution.

Here is another example for a task scenario that provides too much information:

> You are a student in a 19th-century English literature class. You need to find a book by Nathaniel Hawthorne, an American novelist who was influential in the Romantic movement, and who had resided in Salem, Massachusetts. Click on the catalog search at the top of the mobile website to find the book and search by title.

This example has too much context, providing extraneous and irrelevant details that have no bearing on the task's completion. Would the user utilize the author's name to find the book? If so, then why did the task end with finding the book by title? This example also leads the participant, as they are given specific directions on what to "click." This is also problematic since on most mobile websites and apps, the user cannot click anything.

Generally speaking, task-testing scenarios should also have an "end" to the task. Tasks that have a correct solution are useful to record the percentage of successful completion rates. In this example, the end of the task is whether the user found the book *Frankenstein* or not. Not only does this indicate a task success or failure, but it also prompts the participant as to whether or not it was or wasn't successfully accomplished. It also signals the proctor to move on to the next task.

On the same note of having an end to a task, it's worthwhile to write out the "solution" to a task scenario for comparison. This includes the predicted navigation paths a user may take to successfully complete the task. A task may also not end if the user encounters an error or if they choose not to complete the task, which can be considered a failure.

Tasks that have no end, or open-ended tasks, have less direction for the user and may be useful for exploratory purposes. These tasks can also be used to collect feedback. These types of tasks may be better suited for initial development or prototypes. Due to their inherent nature, success rates cannot be applied to tasks that have no end.

When writing scenarios, they should be presented using accessible, natural language. This is especially true for library-related testing due to the jargon that is commonly used. Unless the audience is librarians, terms such as *stacks*, *periodicals*, and *serials* should be avoided, since they can confuse the participant. Examine the above examples for our *Frankenstein* search, and see that the word *catalog* is not used. This wording is intentional. This can also apply to unfamiliar words or titles. Unless your intended audiences are practitioners of a foreign language, it may not be a good idea to have participants search for something in a different language.

Regardless if the interface can be used, the scenario should ensure that the task set before the user is completely understood. A poorly constructed task misdirects a user or may not be useful for testing the correct scope of the project. The articulation of the task is important, as it can affect the validity and reliability of the participant feedback. To accommodate users, it may even be a good idea to have task scenarios written out, rather than having the proctor reading them to users. This allows users to go back to a task if they misheard the proctor reading it.

This is an example of a poorly worded task scenario:

> Find the monograph "Diarios de Motocicleta" in the reserve stacks.

There are a number of problems with this task scenario. The participant is given rich, library jargon in order to complete the task, which shouldn't be necessary. *Monograph* and *reserve stacks* are not words that are typically used among casual library patrons, nor would they be commonly seen on a library website. Also, the title is in another language, which may confuse users who may not speak Spanish. Most importantly, this task gives neither specific directions nor a specific context to guide the user on what to do during the usability test session. Does this task refer to using the library catalog or the library site search? If the user does not know what the word *monograph* means, perhaps they would search for that term instead of the title of the book. It also seems that the title of the book is not explicit, which can further complicate the task scenario and skew results.

Task scenarios should also be written so that they are not leading the user. Providing step-by-step instructions is not appropriate for a usability test. The general idea is to observe how users utilize a site without any given explanation or instructions. For example, it would be unwise to indicate what to tap, drag, or swipe on a mobile website. This could lead to tainted and unusable results. Answering questions during the task test is okay, just so long as it doesn't direct the user. At times, even not answering questions during task testing may be the best approach, so that the proctor does not accidentally lead the user.

Open-ended follow-up questions can be used immediately after a participant attempts a task scenario. Asking open-ended questions can give insight into this experience. At times, it can provide more information that may not be available within the context of the task scenario. This can be as simple as asking what the user thought of the experience. For the task scenario where the participant was asked to find *Frankenstein,* the follow-up question could be "What did you think about trying to find that book?" This feedback can be about the process of the task, which task writers may not initially conceive of. During one usability testing session, I had asked what a student thought about the process of finding the library's hours on a mobile site, and he went off-topic and discussed his dislike of the layout of the mobile page. This can be useful, especially if there are other users who share similar opinions.

Task scenarios should generally be independent from one another. Participants should be required to fulfill one task in order to go on to the next one. For instance, imagine a task where a user is asked to search for a book. The next task could require the user to seek the book's call number. If the first task fails, then the user is unable to complete the second one since it is dependent on the first. Independent task scenarios also allow tasks to be randomized between participants.

Once tasks have been written, it may be a good idea to test them on nonparticipants not involved in usability testing. These nonparticipant users may be representative samples of participants. A dry run of task scenarios can help to revise task scenarios, to ensure that they are well-understood, do not use confusing language or jargon, and ensure that they can be successfully completed. A pilot of the usability test can offer improvements before conducting an actual usability test.

Depending on the scope and purpose of scenario tasks, it may be necessary to establish if tasks should always start at the same point in the system. This is also determined by the order of tasks the participant must undertake. For instance, if the usability tests focus specifically on the home page or starting screen, then that would be the initial starting point for task scenarios. Sequential tasks would follow at the same starting point. If the focus is broader and examines the system as a whole, then tasks should be written so they can be attempted on any page or screen within the mobile project.

For mobile users, task scenarios are a little different because of their mobile context. The task in this case could easily be changed to include such information:

> You are a student in an English writing class. You need to find an e-book version of the book *Frankenstein* for your smartphone or tablet computer. How would you find this book on the mobile library website?

Here, we have altered the scenario specifically for mobile devices. An e-book is also used in this example since these types of resources can be viewed on mobile devices. If there are specific-device users that need to be tested, the example task scenario can be rewritten to omit unused devices. Later in this chapter, the idea of hybrid lab and field-testing will be explored further, using the mobile context within a task scenario.

After the task scenarios have been written, pre-tested, and revised, the next step is to recruit our participants. From there, we can configure the hardware, set up the testing space, identify how to record the test, and formulate test scripts. Once that is determined, we can begin to recruit participants for the test.

RECRUITING

Participants for your task testing should ideally be potential users of your mobile site. For public libraries, this may mean a wide variety of users with an even wider variety of experience with mobile technology. In an academic library this may include students, faculty, and the general public. It may also be necessary to seek approval from the institutional review board to permit testing human subjects, depending on your institution. Recruitment can be publicized in a wide variety of ways (see the next section on advertising), but participants shouldn't be internal developers or users.

Typically, the standby sample size of a usability test is five participants. This is based on the work of Nielsen who argues that after more than five users the test reveals less usability problems and instead revisits the same problems the previous four users have already discovered.[2] The logic is that from 0 users, you learn of no usability problems; from 1 user, you find some problems; after 2 users, there are repeated problems from the first user; and from 3 users, there appear the same problems as the first and second. After that many users, you start seeing the same repeating problems, where there is no point in seeing the same problems over and over. Nielsen argues that it's appropriate to have more users when there are more distinct audiences of the system, such as a system intended for children and adults.

Faulkner argues that more usability problems can be revealed with samples as large as 10 or 15.[3] Nielsen's five-user sample study is dated, and a sample of that size may not accurately reflect the target population. Nielsen's counter-argument is that instead of increasing the number of subjects, it's better to increase the number of tasks to test, or test subsets of specific website features.[4] Faulkner's point of view is intended for testing for statistical purposes and not users' insights. Participant samples may be further constrained based on the resources available and possibly the stage of development the mobile project is in. I believe it's best to have a small number of participants if the mobile web development stems from a mobile-first or responsive design perspective. These philosophies, after all, argue for less sophisticated navigation and design and stripped-down content and limited features, in which case most problems can probably be found from a smaller sample.

Incentives are helpful rewards used to recruit users. Incentives could be cash, gift cards, free giveaways, or even food. What is important is to have participants show up for the usability test. Some participants may not know what they are signing up for and care only for the reward. An appropriate incentive depends on the type of test that is being used. Longer, in-person tests or multiple types of testing may warrant a cash incentive or a gift card, whereas remote testing may suggest a free giveaway or electronic compensation. Library-branded items are an excellent idea for a giveaway. Not only does this provide something for your user, but it promotes your institution as well. Incentives, much like the usability study and use of participants, may also be under review from the institutional review board.

ADVERTISING

Once you've decided on an audience, the next step is to advertise your usability test. Flyers, e-mail, social media posts, and cold calling can get the word out about your usability test (see figure 5.3). If you are using the Web, an online form can be used to screen applicants for specific devices. While advertising for participants,

FIGURE 5.3
Example flyer advertising a usability study.

remember to include incentives, how long the testing will take place, whether a device is being provided or brought, and contact information. Another way to recruit users is through in-person recruitment. This involves approaching users in person and asking if they would be interested in participating in a usability study. This may be difficult if users of a particular device are sought, yet it may seem worthwhile to approach users if they are seen using it. Also note any institutional rules and procedures that may review the recruitment of human subjects.

HARDWARE

Mobile website design and mobile usability share the same challenge: the variety of devices that are used. How is it possible to test the different types of tablets, smartphones, e-readers, and other mobile devices that are out there? For initial task-scenario testing, it may be best to use a paper prototype or desktop emulators.

This is a good way of testing the waters, to gain some insight into how users may initially use the device.

A paper prototype could be as simple as a drawing of a mobile site or app on a piece of paper. This is useful for the initial design stage of a project because it is fast, inexpensive, and relatively easy to implement. The idea is to draw out a mobile interface and gauge users' reactions to it. This is helpful in the early planning stages of a mobile website, and can provide a multitude of information, such as user reactions, opinions, and expectations. This may also improve the workflow and production of the website since potential problems are identified early on.

A desktop emulator simulates a mobile device on a desktop machine. The Google browser Chrome has a built-in emulator in its developer tools window. Screenfly (http://quirktools.com/screenfly/) is a browser-based screen emulator that allows a website to be tested in various devices. This type of usability testing helps identify problems before launching the mobile project.

Despite the advantages of paper prototyping and desktop emulation, there are downsides. Paper prototype and desktop emulation do not provide the same tactile feedback or response of using an actual device. Participants may also not uncover all the usability problems of the mobile project. It is also unnatural for users who may have had prior experience in using mobile devices. Picking up a sheet of paper with a drawing may not be an adequate substitute for a mobile web page.

The final deliverable may be best tested on a real, actual device. This doesn't necessarily mean that your institution go out and purchase every known device that uses your website. Instead, use the most used or top three most-used devices for your users to test on. The device does not necessarily need to be the most current piece of hardware. Instead, use analytics or survey information to determine what mobile device visits the library's website the most.

Another solution for testing your mobile website is to have users bring their own device(s), or BYOD. Not only does this free you from purchasing and providing a device, but BYOD allows the user to utilize what they are most comfortable with and what they use regularly. This way, usability issues are focused on the mobile project and not on the device. BYOD may also be used to set the foundation for a field test type of situation (more on that later).

BYOD does come with a number of disadvantages. First, you would need to screen users for particular devices, which may hinder recruitment efforts. Boundaries and screenings for devices may be necessary, to ensure that devices reflect those used by intended audiences. For instance, it might be best to avoid having five users utilizing the same tablet. It wouldn't make a lot of sense to test your mobile site with a ten-year-old flip-phone with web capabilities if there is only one user visiting the site with it. Users should also be reminded to charge their devices before usability testing takes place. Participants may also need to be compensated if they are using their own data plan and if they are unable to use or have access to Wi-Fi.

Instead of participants bringing in their own devices, providing devices to test has its own advantages and disadvantages. One merit is having more control over the device. Proctors can make sure that their device is connecting to Wi-Fi and that it is fully charged for the test. It's also important for participants to use a device they are familiar with. For one usability test I conducted, I used iPads for testing. The objective was to specifically examine tablets and mobile websites. However, there were some users that were unfamiliar with touch interfaces. It took them a few task scenarios to become comfortable with using these tablets.

Regardless, if users are bringing their own devices or they are provided with them, it's important to get the appropriate ratio of devices that are used on your site. Pendell and Bowman of Portland State University used the current market share of devices, with 4 iPhone users, 4 Android users, and 4 Windows or Blackberry phones for their BYOD mobile usability test.[5] Website log files or analytics can also be used to determine the proper sample percentage of what mobile devices are being used. For a future usability test, I plan on using the top most-used devices that visited our mobile and desktop sites. This information can be quickly gathered through Google Analytics or Piwik. Again, the objective of finding participants is to have those devices that are reflective of the intended audience.

TESTING SPACE

The testing space of a usability test can also be a factor in observing usability. Typically, there are two types of usability testing: laboratory and field-testing. Each of these testing spaces has its own merits and faults within the context of mobile device testing. First, let's examine a typical laboratory testing structure.

Don't think of laboratory testing as a sterile, closed-off room. It's actually just a controlled environment that is conducive to usability testing. It could be as simple as an empty conference room, a break room, or a room with a door that can allow for privacy. The participant should be comfortable, free of distractions, and be able to concentrate during the testing session. If your institution is lucky enough, a room with a one-way mirror would be ideal for testing with observers. If remote observation technology, such as cameras or screen capturing, is available, two separate rooms may be used for the usability test. One room will include the participant and proctor(s), and the other room will have observers to watch the test.

A laboratory-type testing space should have the appropriate technology available to support the usability test. Obviously, this includes an active network connection to support Wi-Fi-enabled devices and electricity to charge them in between or during test sessions. A stopwatch can be used to monitor elapsed time during each task. A handheld clicker counter may be useful to track the number of taps a user presses on a handheld device. This may also be automated or reviewed if the sessions are being recorded. Video and audio recording of participants in the testing space is a must. If these facilities are unavailable, it's suggested that an additional proctor or note taker be present to record the results of the testing session.

As mentioned before, observation technology may also be used either in a one-room or a two-room usability test setup. Cameras and screen-capturing software have become more compact, inexpensive, and may even have wireless capabilities, making it easier to observe a usability test remotely. In the observation room, it may be conducive to have multiple monitors to track the screen capture of the test, a camera to observe the user, and another monitor to observe the interactions of the mobile device. Recording these remote sessions, through digital means, may also be useful for future review and comparisons. Again, it may be necessary to seek the appropriate institutional requirements for audio and video recording of participants.

The drawback of a laboratory setting usability test is that it happens in a vacuum. Even though a multitude of factors are controlled, such as Wi-Fi and electrical infrastructure, observation equipment, and remote observation, this doesn't represent a real-world, accurate scenario when it comes to mobile devices. For example, the use of video and audio recording may seem unnatural to the user and may cause anxiety, which may alter the test. The presence of a proctor or observers can also cause discomfort among participants. Despite this, laboratory usability testing, if carried out correctly, can identify problems with your mobile project. Figure 5.4 is an example of a laboratory usability test room setup.

Field-testing sounds exactly as it is: testing that is done in a real-world environment and context. In previous chapters, we examined how the context and portability of mobile devices can be much more involved than using a desktop machine. Field-testing can uncover some of these issues. A proper field-test setting is anywhere a mobile device would be used. This could be a crowded cafeteria, in the library stacks, or in line at the circulation desk. There are other environmental factors that add the context of mobile devices. Dropped network and cell connections, or a potentially loud or distracting environment, allow observing

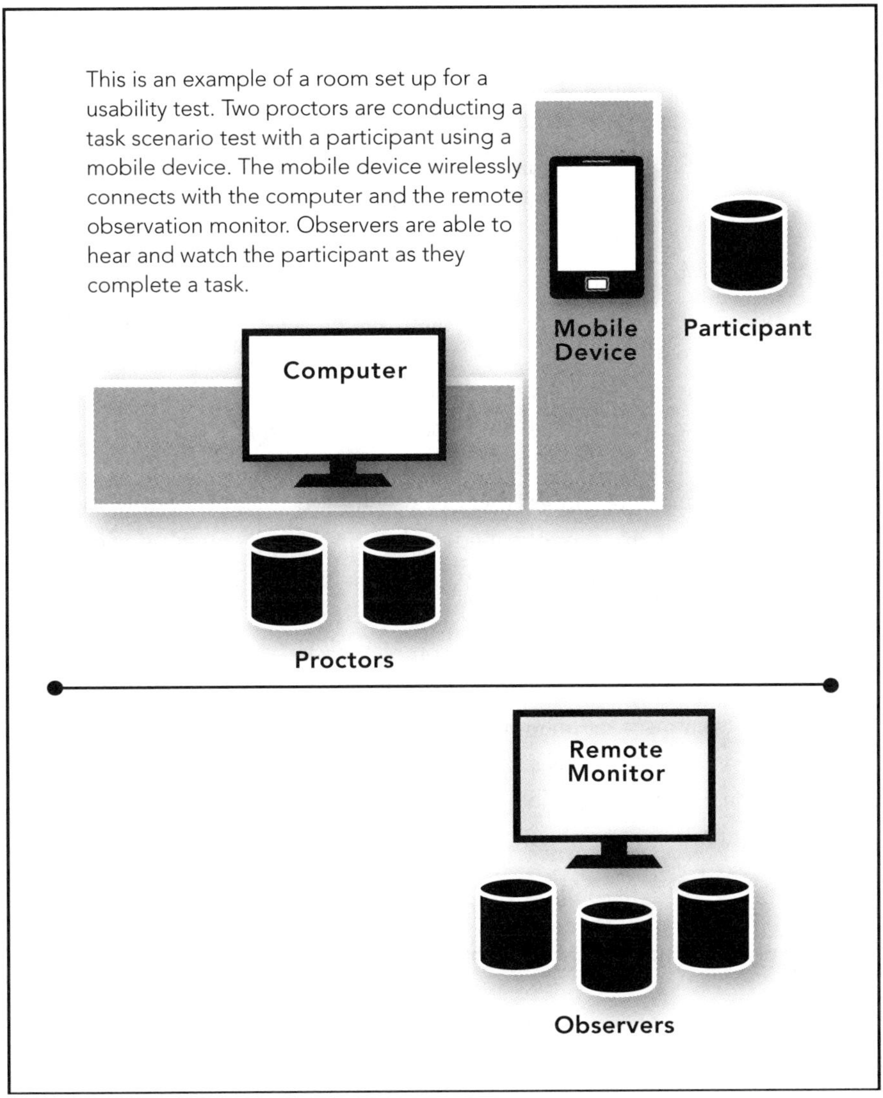

FIGURE 5.4
A sample room setup for a laboratory-type usability study using a mobile device.

participants in a more natural setting. This provides an insight that laboratory testing just cannot provide.

Although more data can be gained from conducting a field test, the distractions and contextual environment can make it more difficult to administer and record the test. Task scenarios and TAP responses may be difficult to hear in a busy environment. Ambient noise and distractions may also make it difficult to audio and video record the testing session.

There are also technological considerations for field-testing mobile web projects. It may be necessary to have the participants wear a microphone headset to record their responses. Field-testing may also hinder the ability to remote screen-cast the participant. It may be possible to download and install screen-recording apps on these devices, such as Rec. (https://play.google.com/store/apps/details?id=com.spectrl.rec), UX Recorder (www.uxrecorder.com/), or Magitest (http://magitest.com/), yet this may require installation on the participant's device or testing device. The downside is that the device must possess enough space to record the user's screen recordings, and be available so files can be downloaded later. The recent Google Glass eyewear may also be incorporated into usability testing. The device is unobtrusive, wireless, and capable of recording video data.

To further complicate things, there is also hybrid usability testing that uses characteristics from both laboratory and field-testing. For instance, users can bring in their own devices into a test setting yet can be remotely recorded by a proctor nearby. A wireless headset, also digitally recorded, can be used to communicate with the participants as they attempt to complete tasks. A more inexpensive solution is to forego the hardware and to incorporate simulated, environmental contexts into testing scenarios. One method is to interrupt users as they attempt a task with a sudden change, such as a hypothetical dropped Wi-Fi connection. This is a pale substitute for an actual field test, yet can be used to gauge how users react to potential mobile contexts.

RECORDING MEASUREMENTS

Usability testing can be recorded in either an analog or digital fashion. Tried and true, pen and paper is an analog type of recording that can be used in usability task testing. This method is the most inexpensive and simplest way to keep notes. It's easier if more than one person is involved with the testing session, such as a proctor to instruct the participant and a note taker to record the session. A task testing scoring sheet can be used to record metric task data (see figure 5.5). This makes task responses organized so they can be easily analyzed after testing. This method is not foolproof, however. Proctors recording responses may accidentally miss or even record inaccurate responses.

Participant ______________________________

	Mouse Clicks/ Taps	Page Views	Time (mm:ss)	Task Success or Failure (S/F)
Task 1				
Task 2				
Task 3				
Task 4				
Task 5				
Task 6				
Task 7				
Task 8				

ERRORS NOTES

FIGURE 5.5
Example of a usability test task matrix that organizes the taps, page views, time, and success rate of individual tasks.

More commonly, video and or audio recording can be used to document a usability testing session. This ensures that nothing is missed when testing participants. Post-test notes can be shared with others for later review. As mentioned, further data can also be collected if you are screen casting (recording the screen) or video recording the participant's face to gather emotional feedback during a task. For initial usability testing, it may be best to use audio recording. Even with this technology, it would be wise to also take notes to document significant events during the testing session.

Below is another example of a scoring sheet used to record an individual testing session (see figure 5.6). The left-hand column includes space for user identification and the metrics for success or failure of a task, time, taps, and errors encountered. This sheet can assist in organizing user feedback for later analysis (see next chapter).

Even though it has become easier and cheaper to record digitally, it is still difficult to accurately record a user interacting with a mobile device. In response, an apparatus called a "usability sled," named due to its design, can be used to record a participant interacting with a mobile device. The sled consists of a platform for the mobile device and an overhead camera. The camera is placed above the device

USER 1: ______	TASK 1: Find a Book	TASK 2: Find an Article	TASK 3: Find Library Hours
Success or Failure	S		
Time	1:06		
Clicks/Taps	4		
Errors encountered	0		

Above is a partially filled usability scoring sheet, with a participant completing the first task. Metrics include the time, taps, and errors that the participant may have encountered.

FIGURE 5.6
An example of an alternate usability scoring sheet which can organize user input.

at an angle where the screen can be recorded. This can be sent to a remote observer viewing room or saved for later viewings. It is typically less obtrusive than an over-the-shoulder camera or a device attached to the participant. These sleds can be purchased commercially or made in the "do it yourself" fashion.

Video recording on mobile devices can be a challenge. Recording a screen while a user interacts with it can be difficult. Lighting and glare can be a problem while recording a screen. As mentioned before, there are a number of apps that can be used to record screen casts, as well as usability sleds. If these resources are unavailable, there are other ways to video record a mobile usability testing session. In a more controlled laboratory setting, a document camera can be used to record the user's interactions with the device. It may take a few test runs to find the best angle and lighting arrangement. Recording the usability test with a webcam focused on the user can also be used to record their reactions. If video recording is unavailable, a microphone and audio recording software on a computer or mobile device will be sufficient to collect participant feedback.

Another way to keep track of user input is to employ software that supports the "mirroring" of mobile devices. Mirroring software allows the screen of a device to be transmitted to a computer either through a wired or, more common, wirelessly over a Wi-Fi network. This is easier to implement under lab conditions on the proctor's own device, rather than on a participant's device in the field. Mirroring a device allows testers to record screen casts of the device connected through a wired or wireless connection. This can be used for later review. Specialized UX mirroring software can also track the user's facial expressions, eye tracking, and finger taps. Combining these two can give a recorded, visual reaction of the user as well as how the website or app is being used.

TASK TEST SCRIPTING

Scripts not only make conducting a test easier, but they ensure that every participant is undergoing identical testing situations. Scripts are usually structured with an introduction, an outline of what the participant is doing, background information, consent form(s), surveys, the actual task testing script, and a post-test debriefing. Scripts are also useful if the person or team who wrote the usability test is not actually proctoring the session. They are also useful for prompting the proctor to ask questions at given times throughout the session. Without a script, usability testing can be varied and unreliable. At worst, proctors may go off track and reveal or conceal the information necessary for regarding a task. It may also create confusion with the participant. Here is the following breakdown for a usability task-testing script.

Introduction

The proctor introduces herself to the participant and what they are doing for that session. The proctor informs the participant what is being tested and what they are going to attempt to accomplish.

> Sample Introduction:
>
> Hi [participant name], my name is [proctor's name]. Today we are going to test the library's mobile website. I am going to ask you to complete a series of tasks.

Outline

The outline informs the participant of an overview of the test session, what a usability test consists of, and the importance of how testing is focused on the system and not the user. This should be explicitly stated to make the user more comfortable. It is natural for a participant, especially students, to try and please the proctor over the course of a testing session. The outline also allows the proctor to encourage the participant to use the TAP (think aloud protocol) as the session progresses. TAP may require prompts to be incorporated into task scenarios to remind the participant to say what they are thinking.

The outline also includes a provision where the participant may be allowed to ask questions, but the proctor may not be able to answer them. For instance, a participant may ask to have a task scenario repeated, but the proctor cannot be asked for instructions or directions for the current task. Additionally, this sample outline states to the user that they can stop the task-test at any time at their discretion. The outline also informs the participant the projected time it may take for the usability testing to conclude.

> Sample Outline:
>
> [Participant Name], it is important to know that this is a test of the library's website and not of you or your abilities. As you go through each task, I ask that you speak aloud of what you are thinking. If there are any parts of the session that you do not understand or are difficult, then it's important to state your honest opinion. You may ask questions, but I may not necessarily be able to answer them. If you choose to do so, you may ask to pass and stop at the current task and move on to the next one. Today's session will approximately take 45 minutes to an hour of your time.

Consent Forms

Consent forms are used to protect the participant, usability testing administrators, and proctors. It also gives proctors permission to test participants. For some situations, this is a requirement if audio and video recording is used to document participants. Refer to your institutional review board's guidelines for sample consent forms for human subject testing.

Survey

The usability test may need initial background information about each participant. Here, a pre-test survey can be used. The survey can include questions regarding age, gender, or mobile device use and behavior. For instance, you may want to know how much time mobile users spend online, or what type of mobile users are visiting the library website. The topic of post-test surveys will be revisited in the next chapter.

> Sample Survey Prompt
>
> Before we begin the task-testing phase, I would like you to fill out this quick survey. The information you're giving us will only be available to us. Your name and identity are not included with this data.

Task-Scenario Testing

The task scenarios that you or your usability team has constructed are used at this stage. Recording begins at this phase and the task testing is initiated. Proctors should use paper and pencil, screen-capturing software, and audio and video recording to record the session. This also offers a final opportunity if the user has any questions prior to the actual test.

> Do you have any questions before we begin? If not, let's start.
>
> Imagine that you are a student in an English class and you need to find the book *Carrie* by Stephen King using the library's mobile site on your smartphone. How would you do that?
>
> [Record Results]
>
> What did you think of while performing that task?
>
> [Record Results]
>
> What did you like or dislike?
>
> [Record Results]
>
> Thank you. Let's continue to the next task.
>
> [Begin next task]

Post-Test Debriefing

The end of the task testing is where the session wraps up. The proctor can thank the participant for their time and input. Participants may also receive compensation for the session. This also allows a final opportunity for questions or a post-task testing survey.

Sample Post-Test Debriefing

> That was the last task for this session. Your input will be used to improve the library's mobile website. Please fill out this post-test survey.
>
> [Give participant post-test survey]
>
> Thank you for your time.
>
> [Give compensation to participant]
>
> Do you have any further questions about the session?

CONDUCTING THE TEST

Now that we have our tasks, data collection methods, and scripts, we can finally start testing. Participants will be recruited using the methods above, scheduled for a task-testing session, and perform the session where data will be collected. Proctors will administer the test and record the necessary information about the participant and tasks. This process will have concluded the first round of usability testing.

Usability Testing Checklist

> Testing Script
>
> Consent Form
>
> Survey
>
> Task Script
>
> Recording Equipment
>
> Scoring Sheet
>
> Stopwatch
>
> Post-Test Survey
>
> Participant Incentive

TESTING ANALYSIS

Once the first round of testing is completed, data can be collected and analyzed. Task test analysis should directly address the specific goals that were outlined within the purpose and scope defined at the beginning of the study. There might be other usability problems revealed through the study, but it's more efficient to improve the issues the study was set to examine. This is where generalizations about the problems within the system can be made.

Results from the usability test can be categorized in various levels of priority. If there is a high rate of failure of completion of a particular task, then this should be listed as a *critical* priority. Tasks ranked as *serious* priority could be those that are so problematic for users that they discourage completion. *Minor* tasks are those tasks that are of low priority but do not prevent them from completing the task.

For example, imagine the task scenario is to find and call the phone number for the reference desk on the mobile site. For our usability test, our testing device is a smartphone. However, the phone number on the library's mobile site is unlinked and not using the `<tel>` html tag (see last chapter) the user cannot call the reference desk. Essentially, this task cannot be completed. Based on this wide-scale, critical-level failure, the website developer can go back and update the code for the site. A second round of usability testing using the same task scenario and a similar participant pool (but not the same) can confirm if this has improved the system or not.

Input collected from participants answering open-ended questions can be collected and analyzed. For instance, during a usability test that I conducted for my institution, many students were asked to find a book using the library catalog. Every participant was able to complete the scenario successfully. However, the catalog required the user to tap a button to open a link to access a different page containing the catalog's search field (see figure 5.7). A few students asked if it were possible just to have the catalog search on the home page of the mobile site. After implementing this, the time that it took students to find the catalog decreased substantially (see figure 5.8). Even though the success rate was 100 percent, changing this feature solely based on open-ended responses had increased the efficiency of participants seeking a book.

Side-by-side comparison of how pre- and post-mobile website versions changed was based on users' open-ended responses from the usability test.

Over the course of this chapter, the usability study for the library mobile project has been defined, planned, and executed. The next chapter will examine what to do after the conclusion of the first round of usability testing.

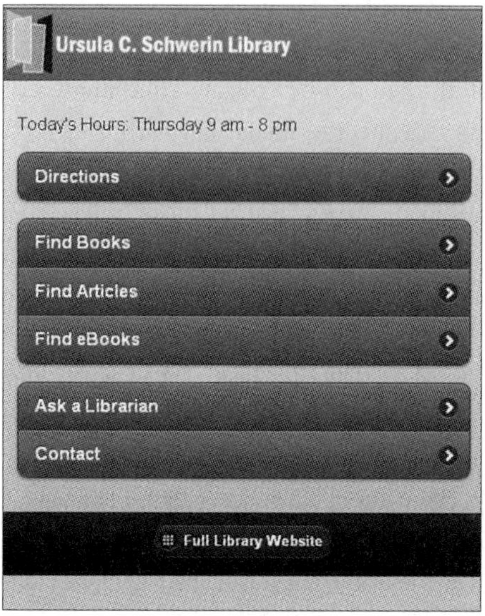

FIGURE 5.7
A screenshot of a library's mobile page prior to usability testing.

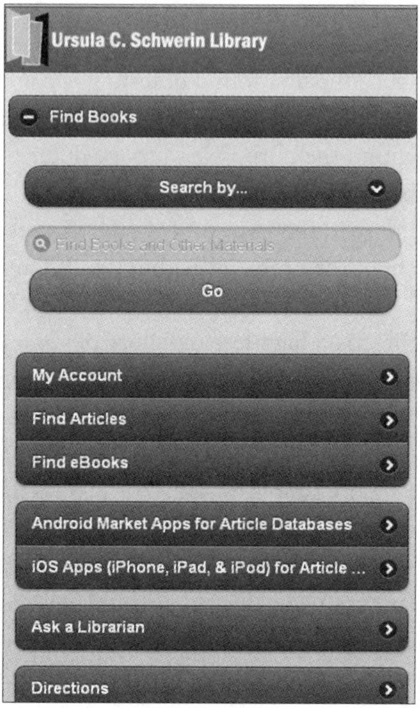

FIGURE 5.8
The modified page post-testing, which had increased the efficiency of participants' tasks.

NOTES

1. Jeffrey Rubin, *Handbook of Usability Testing*, 2nd ed. (Indianapolis, Ind.: Wiley, 2008).
2. Jakob Nielsen, "Why You Only Need to Test with 5 Users," 2000, Nielsen Norman Group, www.nngroup.com/articles/why-you-only-need-to-test-with-5-users/.
3. Laura Faulkner, "Beyond The Five-User Assumption: Benefits of Increased Sample Sizes in Usability Testing," *Behavior Research Methods, Instruments & Computers* 35, no. 3 (2003): 379–83.
4. Jakob Nielsen, "How Many Test Users in a Usability Study?" 2012, Nielsen Norman Group, www.nngroup.com/articles/how-many-test-users/.
5. Kimberly D. Pendell and Michael S. Bowman, "Usability Study of a Library's Mobile Website: An Example from Portland State University," *Information Technology and Libraries* 31, no. 2 (2012), http://ejournals.bc.edu/ojs/index.php/ital/article/view/1913; doi: 10.6017/ital.v31i2.1913.

Chapter Six

Gathering Data after Testing

After the initial usability testing, more opportunities for data gathering are available. Participants can give more feedback before the end of the session. This can be approached as a debriefing process. At this stage of the testing process, proctors can ask participants to fill out a post-testing survey, additional questions about the usability test, or even answer questions that the participant may have.

Typically, debriefing sessions are done immediately after the testing session since it is fresh on the minds of both the proctors and participants. Depending on the type of testing and circumstances of the participants, the post-testing session can be done in the same room as the usability test. In a field situation filled with distractions, it may be necessary for the post-testing to be done elsewhere. Similar to the other testing session, it may be a good idea to video or audio record the post-test session. This will be useful for reviewing the sessions for data collection or sharing sessions with other stakeholders.

SURVEYS

Post-test surveys can be used to collect data on user satisfaction. Likert scales can be useful here, as they are quantifiable and can pinpoint generalities among participants. There can also be an opportunity to ask testing discrepancies. For example, a participant may have completed a task, yet rated it very low on the post-testing survey. The proctor can then ask questions on why they felt this way.

Post-testing can be useful for proctors, for they can ask more questions tailored to the specific participants. For example, a proctor can inquire why one method to complete a task was used instead of another. Open-ended questions could also be asked, such as asking participants what they thought of the website being tested. For mobile usability, this may also be useful to see if they use other platforms other than the one tested on, such as tablets and other devices. These types of questions can also reveal issues that developers, administrators, and the usability team may not have foreseen. It may also be beneficial to ask what task participants perceived as the easiest or most difficult to perform.

SYSTEM USABILITY SCALE

Post-testing can also benefit from a specific survey called the System Usability Scale. The System Usability Scale (SUS) is another method for evaluating the usability of an application. Developed in 1986 by John Brooke, the SUS is a ten-item survey.[1] Each survey question uses a five-point scale, which ranges from strongly disagree to strongly agree. The SUS includes the following:

1. I think that I would like to use this system frequently.
2. I found the system unnecessarily complex.
3. I thought the system was easy to use.
4. I think that I would need the support of a technical person to be able to use this system.
5. I found the various functions in this system were well integrated.
6. I thought there was too much inconsistency in this system.
7. I would imagine that most people would learn to use this system very quickly.
8. I found the system very cumbersome to use.
9. I felt very confident using the system.
10. I needed to learn a lot of things before I could get going with this system.

SUS has been widely used as a valid and reliable instrument for usability studies. According to the U.S. government usability website, it has been referenced in over 1,300 studies.[2] The main drawback of SUS is that the scoring process can be difficult to process. First, each item response should be converted into a number. Odd-numbered SUS items should subtract 1 for that score. Even-numbered SUS item responses should be subtracted from 5. In order to obtain the final score, survey items are added together (sum) and then multiplied by 2.5. The average score of the SUS scale is 68. A higher score is considered above average in usability and a lower score is considered below average. SUS may be useful as supplemental information, but does not point out specific usability problems with a mobile application.

POST-TEST OPEN RESPONSES

Participants should have an opportunity to ask their own questions to the proctor about the experience. This opens the conversation up for potential problems of the site and its usability. Participants may also suggest design changes. Proctors and the usability team may choose to utilize or ignore these comments.

At the conclusion of the usability test, participants should be allowed to reach out to the proctors for further contact. This can occasionally lead to other issues that were not necessarily discussed during the post-test debriefing. This can work both ways, as proctors may need to contact the participant to follow up on other usability issues. It may also be a requirement as part of your institutional review board for participants to know what and how their data has been used.

THE FIRST ROUND OF TESTING IS DONE . . . NOW WHAT?

After the first round of usability testing, the data collected should be assessed and analyzed. This would be the stage of the usability testing process to look for commonalities among participants. One method that can be used is to recognize consistent patterns among testing sessions. There are different ways to identify these patterns. Perhaps one task took a very long time to complete and you are interested in why. Another pattern found in testing could be a crucial task that every participant failed to complete. If your usability sessions utilized a scoring sheet (see previous chapter) and there are several consistencies among participants, then this is probably a pattern that needs immediate investigating (see figure 6.1). Examining and reviewing transcripts of audio and video recordings can be useful in finding patterns among participants' responses. Identifying these problems can help prioritize what should be modified first.

TASK 1: Find a Book	User 1	User 2	User 3
Success or Failure	F	F	F
Time	1:06	2:13	5:00
Clicks/Taps	14	20	26
Errors encountered	10	10	10

In this filled sample task-scenario scoring sheet, all users failed the task of finding a book on the mobile project. The pattern that can be derived is consistent failure among three users, and a large number of taps for an unsuccessful attempt.

FIGURE 6.1
A completed task-scenario scoring sheet, indicating individual success rates. This can be used to find patterns that can lead to usability obstacles.

Various experts have created different scales of severity for usability problems. Nielsen has a five-step scale ranging from 0–4.[3] Rubin has outlined a four-level scale with "unusable" as the most extreme, to severe, moderate, and irritant (least severe).[4] Dumas and Redish have listed 4 levels, where the 1st level prevents task completion, the 2nd creates a significant delay and frustration for the user, the 3rd has a minor effect on usability, and the final and 4th level has subtle and possible enhancements and suggestions.[5] Whatever scale is used, listing problems can help tackle them efficiently and systematically.

Once problems are scaled, they should be repaired in order of the most severe to the least severe. Fatal problems, obviously, or those where a task cannot be completed should be prioritized first. In terms of mobile devices, the process of compiling this post-testing data is not that different from that found in desktop usability testing. However, what should also be considered are the "mobile context"–type problems. For instance, there could be problems successfully completing a task due to the intermittent Wi-Fi connections inherent in mobile devices. Tasks using smaller, touch-enabled input, such as those from virtual keyboards found on newer smartphones, could take longer to complete. Separating such problems from those outside of the direct control of the development team can be easier to repair.

Compiling Data

Quantitative Data

Statistics can easily summarize data into digestible measurements. This type of quantitative data is useful to administrators and developers to improve the system. There are numerous statistics that can be calculated after completing the first round of usability testing. One statistic that can be determined is the success-to-failure ratio. This is the percentage of successes or failures for each task, which can be determined by:

number of success / number of testing sessions × 10 = success percentage

Success percentages should also be given in context. Time, the number of taps, and other metrics are taken into account, so developers and stakeholders can get a clearer picture of what is going on with site interactions. Specific information from open-ended questions, post-testing surveys, and other participant feedback can also be used to provide context.

Imagine a scenario where there is an 80 percent success rate among participants of finding a book using the mobile library's site catalog. This high percentage may indicate that the task can be accomplished successfully among several participants, but what if there are other factors involved? Perhaps it took several minutes for the participant to find a book in the catalog, or there was frustration expressed during the post-testing session. Success percentages do not always tell the whole story when it comes to usability testing.

As mentioned in the last chapter, timing is a metric that can be extrapolated from usability sessions. Measuring how long it takes participants to complete a task provides useful data, which is especially critical for mobile usability. It's important to acknowledge the demands of mobile users, as they require information immediately, and typically on the go. This metric can be used in a variety of ways. The average time, or mean, it takes for participants to complete a task can also be a useful metric as seen in the example above for developers and designers. It should also be noted that TAP responses could affect the timing of a task, so this too should be considered when analyzing results.

sum of participants' time for a given task / number of participants = average time

Timing can also come up with another metric, the range. The range is the completed time of a task taken from all participants, ranked from the lowest participant's time to the highest. This can be used to make comparisons between users. There may be problems with a particular task if there are extreme ranges between participants, or timing can be used to identify anomalous participants or outliers. There could be a single participant whose time was very remote from the range. Again, the mobile context should be taken into account when evaluating the range, since slow connection speeds can hinder the progress of a task.

Some usability studies have used a baseline time for users to reach based on the time of an expert (see previous chapter). This can be incorporated into the range and be used to compare the timings. This may be useful to compare an expert's familiarity with a system with that of participants who may not be familiar with it.

Clicks and taps are another metric where many patterns can emerge. Like the other metrics that usability tests can record, clicks and taps are only one piece of the story and require context. Reducing the number of taps can reflect the arrangement of content, or the information architecture, of the site. A well-structured architecture enforces consistency, making it more intuitive to use. Ideally, the less number of clicks it takes for a user to complete a task the better, yet this isn't a hard-and-fast rule. Taps can be used for exploratory purposes for the user, such as scrolling a web page. Library websites may also have numerous processes that require a large amount of tapping. Logging taps that correspond with an event may be better than counting every individual tap.

Time can also provide a context for taps. If there are a high number of clicks and a large amount of time to successfully complete a task, then this may be a problem area. The recordings of participants as they underwent the TAP can also provide context for a great number of clicks for particular tasks. The same statistics pulled from the time metric—mean and range—can also be applied to clicks and taps.

Qualitative Data

Categorizing similar responses together can be used to summarize participant TAP responses. Attributing these results to individual tasks can create usable information for developers, and as explained earlier, can provide context for quantitative data. Comments that reflect the interface, the process of accomplishing the task, and any obstacles that participants encountered can be used to identify usability problems. If several users note the same problem over and over again, then this would be worthy to report. Post-testing responses can also be categorized and analyzed in a similar fashion. During post-testing, participants may even refer back to specific tasks, which can be grouped together for the final report.

Now that the data has been compiled, this can be presented to developers and stakeholders to make appropriate changes to the library's mobile website. Here is where error severity levels can come into play. The most fatal tasks should be rectified first. This includes any tasks that have a high rate of failure among participants, lengthy taps and increased times, or task attempts that are outright unsuccessful. These tasks are generally easy to identify, but it may be difficult to distinguish them from user errors, system errors, or problems within the mobile context. Analysis of usability errors must be systematically taken apart. If an involved process takes many steps, then it's important to evaluate each stage of the process.

For example, imagine a user is tasked with making an interlibrary loan request on your mobile library app on their smartphone. Let's also imagine that the user must find the interlibrary loan section from the application's home screen. After finding this resource, they typically enter an authentication log-in, and from there make a request for their loan. This can take several steps, with an error, or even multiple errors, that could occur in any one of them. Perhaps they entered the wrong credentials to enter the interlibrary loan system. Maybe they didn't type in the correct title for their loan request. The error may even be at the very beginning of attempting the task, such as finding the link on the opening page of the application.

Accurately finding and identifying these errors are crucial for developers so they can repair the underlying problem. Again, this must be placed within the context of the mobile user, including their experience with the system, the performance of other tasks during the session, and the device/platform they are using.

After the problems, errors, and participant responses have been compiled, analyzed, and placed into context, the next step is to inform developers and stakeholders. This can be done informally or formally, depending on who your audience is. It may be best to assume that this report will be written specifically for those not directly involved with usability testing. This may include:

- Participant backgrounds
- Tasks performed
- Device(s) the test was conducted on
- Description of the testing environment
- Data analysis
- Issues
- Recommendations for change

Participant Backgrounds

Participant backgrounds are just that. This is information about participants that may be useful for developers to know who have tested their system. This can include demographic information such as education background, age, and ethnicity. Education background may be useful for academic institutions to compare specific user groups, such as incoming freshmen and final-year students. Having a wide range of ages and ethnicities also ensures that there is a wider presence of voices using your system. Other participant information may also include collected data from pre- and post-test questionnaires and surveys. It could also be worthwhile to consider other observers' notes, and if they witnessed any specific characteristics of participants during the usability test.

Personas can be developed out of participant backgrounds. A persona is a representation of a project's intended user, with specific behavior and attitude characteristics. These can be used to put a "face" on users so that developers can relate to them more. Usability teams can construct personas based on the participants from user research and usability testing. The small sample of participants from the first round of usability tests may not be enough to generate many personas, but common characteristics can still be used. Analytics information and surveys can also be used to supplement persona templates.

Tasks Performed

Developers and stakeholders should know the specific task scenarios that participants were asked to perform. It wouldn't make much sense to give a list of errors to someone if they don't understand what participants were attempting to perform. It makes more sense to developers to see the system from the user's perspective. It may provide further context to include pulled quotes from participants' TAP responses for specific tasks. Depending on the dependence between tasks, it may be necessary to include the order in which tasks were performed.

Device Testing

This report should also include the types of devices that the usability tasks were performed on. Each device has its unique qualities that differentiate themselves from each task test. For smartphones and mobile devices, it may also be important to pinpoint how these devices connected to the network, either from Wi-Fi or cellular data connections. Other information such as screen size and resolution would also be helpful. This can also be included within the description of the testing environment.

Testing Environment

In a usability report, a short description of the testing environment can provide a clearer picture for developers. Be it a lab or a field setting, the testing environment will give further context of the perspective of the end-user. This may also list out the specific ways data was recorded (video, audio, or both), and the process of in-person observation if there was any. A description of how the testing environment was arranged can be useful as well.

Data Analysis

The heart of the report will be the data analysis section. This includes a summarized analysis of qualitative and quantitative data of the usability test. Specific responses for each task during the think-aloud protocol or during the post-testing debrief should be included here. Statistics, graphs, and charts can be used to make information easier to scan. Success percentages should be included for each task. It should also be noted if participants had specific responses or questions during the usability test. These can be pulled from TAP responses, or if they were asked directly to the proctor during the test.

Issues

The issues found in the usability testing can include a breakdown of problems with the system. The most fatal flaws should be presented first, followed by tasks in descending order of severity. Other system errors can also be listed for developers to repair. Notable user errors can be included in this section of the report. It should also be noted if participants encounter the same errors within the system. Again, pulled TAP responses can be useful here if they are given within the context of the issue.

Recommendations

The final section of the report includes recommendations from the usability proctors. Depending on the tasks evaluated, this can range from being very simple recommendations to in-depth, step-by-step processes that developers can use to make the system more usable for end-users. These recommendations should be backed up from the data gathered from participants during the usability test.

For example, perhaps all of the participants found it difficult to traverse through the mobile library website to find branch hours on the home page. One recommendation is to place those hours, or a link to the specific hours page, directly on the header of the home page or even on every page of the site. Other recommendations can be more clear-cut, such as better wording for navigation, appropriate feedback, or larger fonts for content and links for mobile devices. These suggestions should be backed up based on the data collected during testing. It may also be worthwhile to include screenshots of some of the screens that participants had seen to explain abstract concepts.

Sometimes all the recommendations from the usability test may not be implemented. Unforeseen circumstances, budget restraints, or time restraints can prevent changes from being made. Instead, what may be best is to address the most severe problems first to ensure a fully working system. Minor and cosmetic problems can be addressed later. What is important is to have enough turn-around for a second round of testing. Library websites may have several other third-party resources connected to it that are live and still ongoing. This includes integrated library systems, electronic databases, discovery tools, and other technologies that are ever-changing and live. Waiting too long to release an updated version of your mobile application may result in a version too dated for modern devices.

Second Round of Testing

The second round of task scenario testing is initiated after the modifications to the system have been implemented. This allows the usability team to see how successful the suggestions and recommendations from the first round are. During the second round of testing, the site is tested again using the same script, tasks, testing environment, and post-testing materials. It would also be ideal to use similar participants, yet not the same ones from the first round. This is to avoid a high usability score from participants who have had previous experience with the application. The participants should also be given the same incentives as the first round of participants. This reiterative testing can be used to validate or invalidate changes that have been made.

A second round of testing isn't always absolutely necessary. Having an initial usability test can be useful just to see how well the site is working with a group of users. A single round of usability can also be used for prototypes and demo applications. The goals of your usability test can dictate whether or not multiple rounds are necessary. However, as mobile technology continues to grow and evolve, there are few reasons not to have multiple rounds of testing.

Continual Testing

Continual usability testing aligns with *iterative design,* where a system is tested, modified, and then retested again and again (see figure 6.2). An initial usability test can find problems with the design, yet continual testing can reveal even more. Madrigal and McClain make the point that severe usability problems can obscure underlying problems.[6] Once the bigger problems are taken care of, fixing the subsequent smaller problems can lead to a more pleasing user experience. Nielsen has noted that usability between design changes can improve 38 percent per iteration.[7]

In a world of varied mobile platforms, interfaces, devices, and users, it is almost necessary to consistently test designs. Unpredictable issues and new problems may

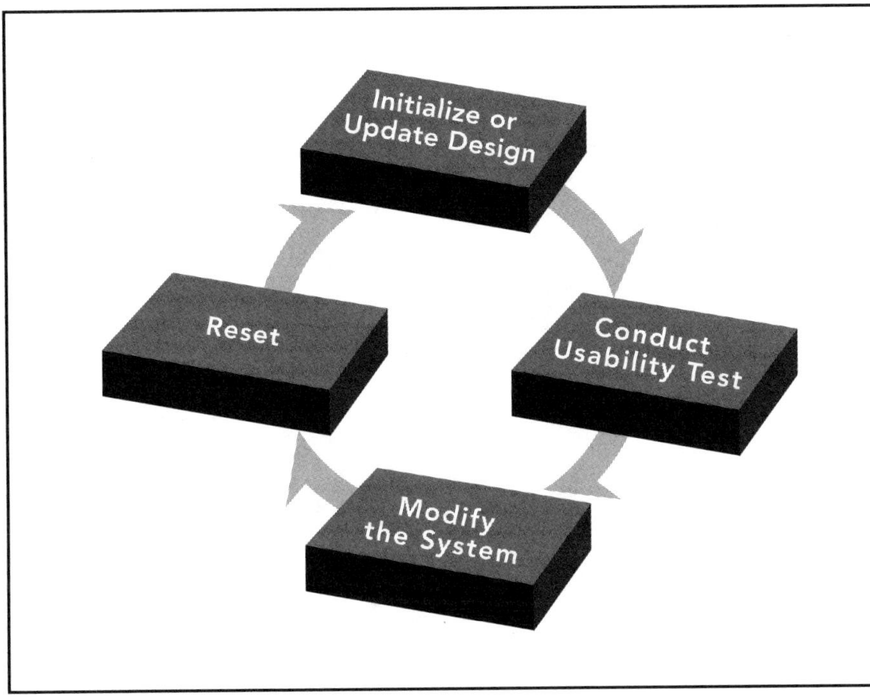

FIGURE 6.2
Usability testing stages.

arise as technology changes. The way that users interact with technology may also alter as time goes by. Mobile devices have changed how users consume information. Libraries must adapt to these new technological changes, to fulfill our mission to provide information access to all.

NOTES

1. John Brooke, "System Usability Scale (SUS): A Quick-and-Dirty Method of System Evaluation," Digital Equipment Co. Ltd. (1986).
2. Usability.gov, "System Usability Scale (SUS)," www.usability.gov/how-to-and-tools/methods/system-usability-scale.html.
3. Jakob Nielsen, "Severity Ratings for Usability Problems," 1995, Nielsen Norman Group, www.nngroup.com/articles/how-to-rate-the-severity-of-usability-problems.
4. Jeffrey Rubin, *Handbook of Usability Testing*, 2nd ed. (Indianapolis, Ind.: Wiley, 2008).
5. Joseph S. Dumas and Janice C. Redish, *A Practical Guide to Usability Testing* (Portland, OR: Intellect Books, 1999).

6. Demetrius Madrigal and Bryan McClain, "Do's and Don'ts of Usability Testing," 2010, UX Matters, www.uxmatters.com/mt/archives/2010/03/dos-and-donts-of-usability-testing.php.

7. Jakob Nielsen, "Parallel & Iterative Design + Competitive Testing = High Usability," 2011, Nielsen Norman Group, www.nngroup.com/articles/parallel-and-iterative-design.

Appendix

Sample Scripts and Forms

Verbal or E-Mail Recruitment Script

Hi, my name is [Your Name].

I am the [Position] Librarian here at [Name of Library/Institution].

I would like to invite you to participate in a research study I am conducting.

As a participant, you will be asked to commit to a 45-minute session.

There are no risks in participating in this study.

You will be given a [Insert incentive].

Any information you provide will remain confidential. The study is completely voluntary.

If you would like to participate go to this URL [Insert screening URL] or e-mail me at [e-mail].

Sample Consent Form

You are invited to participate in a usability test with the library's mobile website. The purpose of this study is to determine how useful the library website is for mobile users.

During the test you will be asked to:

- Fill out a pre-test survey
- Perform tasks on the provided mobile devices
- Fill out a post-test survey.

The test will take approximately 30–45 minutes of your time. At the end of the test, you will be compensated with ________________.

Your participation in this study is voluntary, and you may decide not to participate. If you decide to leave the study, please contact the administrator of the test that you no longer wish to participate.

All information will remain strictly confidential. Data from this mobile usability test will only be available to librarians and the developers of the mobile project.

If you have any questions, please contact the administrator at admin@yourlibrary.org.

By signing the consent form, you give us permission to digitally video and sound record the usability test session.

Name: ____________________________________

Date: ____________________________________

Sample Mobile Usability Script

Hello [Participant Name]. My name is [Name of Proctor].

Today, I'm going to ask you to use the mobile site using this [Name and Type of Mobile Device or User's Own Mobile Device]. Your test will help us improve the [Name of Library]'s [Mobile Project]

I am going to ask that you think aloud while we conduct the test session today. I want you to tell me what you're thinking as we go through the session. You will not hurt anyone's feelings, so feel free to speak what is on your mind and give your honest reaction. I will periodically ask you what you are thinking as well as any follow-up questions to the task.

If you have any questions, feel free to ask them. I may not be able to answer them, as it's very important to see how you use the mobile site without anyone beside you.

At any time you feel uncomfortable with the task at hand, we can take a break or move on to the next task. There is no penalty for you to skip or stop the task.

This is the consent form so we can begin the test. After you are done reading it, please sign and date it.

(Hand Participant Consent Form and Pen)

Once you have signed the consent form, we can begin to conduct the test.

(After Consent Form Has Been Signed Begin Recording)

If you are ready to get started, we can begin with the first task.

Sample Mobile Tasks Scenarios

1. You are a new user and have not used the library mobile website to find books before. Using your mobile device, find the library's online catalog.
2. You are a freshman in an English 101 course. You are asked to find a book on an American author for a research paper. Using the mobile library website, find a book by the author Mark Twain.
3. The psychology course you are taking requires you to write a research paper on child development. You need to find an article for a research topic using an electronic database. Using the mobile website, find the electronic database where you can find research articles on child development.
4. There is an Android app on Google Play for the EBSCO Academic Search Premiere electronic database. You need the database to find information for a history project. Using the library mobile site, how would you access these apps?
5. It is late and the library is closed. You need help with finding information and decide to e-mail a librarian who is a subject expert on what you are studying. You need to ask a librarian a question over e-mail. How would you find the librarian's e-mail using the mobile library website?

Bibliography

David, Matthew. *HTML5 Mobile Websites*. Abingdon, UK: Focal, 2012.

Frost, Brad. "Creating a Mobile-First Responsive Web Design." Html5 Rocks. www.html5rocks.com/en/mobile/responsivedesign.

Griggs, Kim, Laurie M. Bridges, and Hannah Gascho Rempel. "library/mobile: Tips on Designing and Developing Mobile Web Sites." code4Lib 8 (2009). http://journal.code4lib.org/articles/2055.

Lang, Tania. "Eight Lessons in Mobile Usability Testing." UX Mag. http://uxmag.com/articles/eight-lessons-in-mobile-usability-testing.

Weyl, Estelle. *Mobile HTML5*. Sebastopol, CA: O'Reilly Media, 2014.

Index

D

E

F

R

S